59th Annual Symposium of the Society of Experimental Test Pilots (SETP 2015)

Anaheim, California, USA
23-26 September 2015

ISBN: 978-1-5108-1639-8

Printed from e-media with permission by:

Curran Associates, Inc.
57 Morehouse Lane
Red Hook, NY 12571

Some format issues inherent in the e-media version may also appear in this print version.

Printed by Curran Associates, Inc. (2016)

For permission requests, please contact Society of Experimental Test Pilots
at the address below.

Society of Experimental Test Pilots
P.O. Box 986
Lancaster, CA 93584-0986
USA

Phone: 661.942.9574
Fax: 661.940.0398

Setp@setp.org

Additional copies of this publication are available from:

Curran Associates, Inc.
57 Morehouse Lane
Red Hook, NY 12571 USA
Phone: 845-758-0400
Fax: 845-758-2634
Email: curran@proceedings.com
Web: www.proceedings.com

TABLE OF CONTENTS

Society of Experimental Test Pilots
Annual Symposium
Anaheim CA, 23-26 September 2015

Active Flutter Suppression Testing on the X-56A Multi-Utility Technology Test-bed Aircraft

Edward L. Burnett and Michael Hartenstine (AM)
Lockheed Martin Aeronautics Company, Palmdale California

The desire for increased aircraft performance - to fly higher, farther, faster, cheaper, has not diminished, however many of the processes used to achieve these goals in the past have run their course. In order to continue to push performance, engineers are looking at integrating multiple traditional disciplines to leverage the combined capability. Much as the combination of aerodynamics, stability and control with advanced feedback controls allowed the reduced static stability on the F-16 to increase maneuverability, we now seek to extend the techniques of feedback control to allow reduced structural stability. Active structural modal control or "flutter suppression" is being tested on the X-56A by Lockheed Martin under a contract from the Air Force Research Labs. This paper will discuss the history of this technology, and the lessons learned from the engineering, and pilot's developmental test perspective, in fielding this capability on the X-56A. The paper will also discuss questions that remain as this research is continued.

Nomenclature

b	Wingspan
AR	Aspect Ratio
Abbreviations	
6DOF	Six Degree of Freedom
AFRL	Air Force Research Laboratory
BFF	Body Freedom Flutter
CCV	Control Configured Vehicle
CRAD	Contracted Research and Development
FCS	Flight Control System
FMC	Flutter Mode Control
FMET	Failure Modes and Effects Testing
GCS	Ground Control Station
GVT	Ground Vibration Test
HALE	High Altitude Long Endurance
IFAC	Integrated Flight and Aeroelastic Control
IRAD	Independent Research and Development
LCO	Limit Cycle Oscillation
LM	Lockheed Martin Corporation
MAD	Multi-utility Aeroelastic Demonstrator
MFMC	Multiple Flutter Mode Control
MUTT	Multi-Utility Technology Test-bed
NASA	National Aeronautics and Space Administration
NDOF	Number (normally greater than six) Degree of Freedom
RFC	Rigid Flex Coupled
RPO	Residual Pitch Oscillation
UAV	Unmanned Aerial Vehicle
VFG	Velocity, Frequency, and Damping
X-56A	Lockheed Martin Design and Built Research Aircraft

Introduction

In the opening of the movie "The Right Stuff", Levon Helms gives voice to the late Jack Ridley warning us of "a Demon that lives in the air." However, the Demon that is the sound barrier is not the only creature that lurks in the sky. There is also the Dragon that is flutter.

Figure 1; The rocket powered Bell X-1 in flight

Just as the speed of sound presented an obstacle to greater performance, flutter stands in the way of our desire to fly Higher, Further, Faster, Cheaper because our current solution in its simplest form, is to add weight to add stiffness.[1][2][3] Like the development of the X-1 (Shown in Figure 1) was to understand the phenomenon of the speed of sound and to develop technologies to remove it as a limit to performance, we are now working to understand and develop more efficient methods to combat flutter.

History

In 2005 a group of Lockheed Martin engineers requested Independent Research funds to better understand the problem.[4] The group designed and built 5 expendable remotely piloted aircraft for the initial research. The vehicles were simple being built using mostly hobby store components, but were also quite sophisticated. The aircraft had an on-board computer, integrated INS/GPS, structural sensors, air data transducers, and an on-board data recording capability. The vehicles known as BFFs for Body Freedom Flutter, were structurally designed to be strong yet exhibit flutter at a relatively low speed.

Body Freedom Flutter is a coupling of the vehicle's rigid aerodynamic short period and the symmetric wing first bending mode of the flexible structure. Figure 2 shows one of the BFF vehicles in fully developed Body Freedom Flutter.

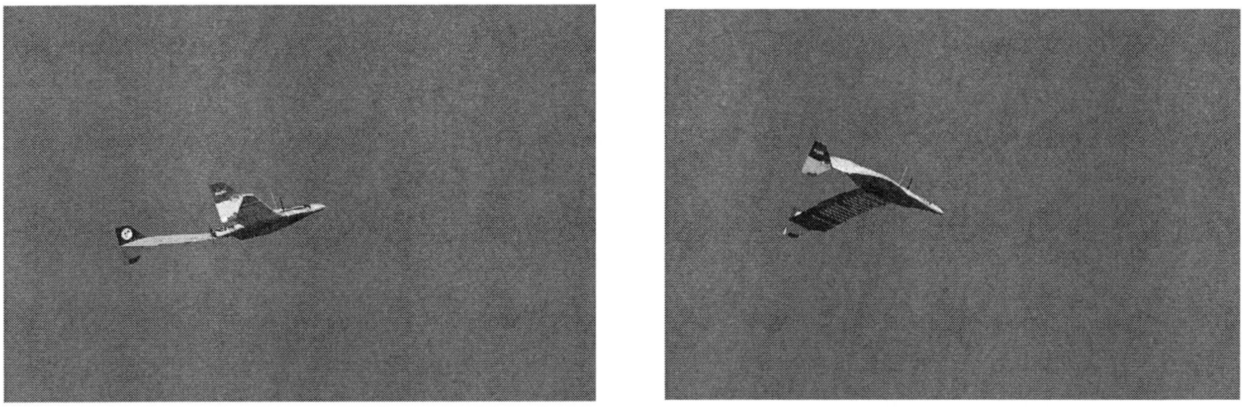

Figure 2; Fully Developed Body Freedom Flutter exhibited in expendable UAV

The first two vehicles were used for open loop model validation. The results from flight test were within only a few percent of the predicted onset speeds and the frequency of the mode was likewise within only a few percent error. These flight test results showed that our modeling techniques were accurate enough to allow for the development of an active feedback control suppression system.

Figure 3 shows a Velocity Frequency Damping plot often called a VFG plot of the BFF vehicle. The plot shows that the coupled BFF mode would go unstable at 46 knots.

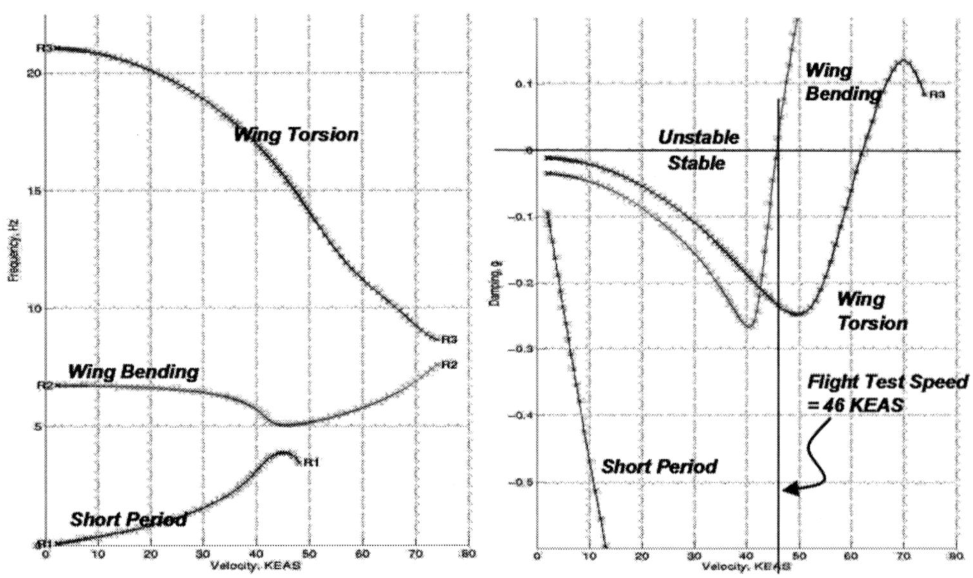

Figure 3; Frequency and Damping predictions as a function of velocity validated aeroelastic modeling techniques

Using lessons learned from the first attempt at suppression of BFF conducted on vehicle number 3; vehicle BFF04 was flown to at a speed 39 % above the onset flutter speed. By the flight test of vehicle number 7 three flutter modes were suppressed simultaneously and the aircraft was flown to an air speed 27 knots or 66 % above the first unstable flutter mode. (5)

The phenomenon of Body Freedom Flutter is not limited to high aspect ratio low tail volume aircraft such as a High Altitude Long Endurance aircraft like AFRLs Sensorcraft concept. (6) BFF can also limit the performance of aircraft designed for high speed. The phenomenon is caused by the coupling of the aircrafts short period and fuselage bending and wing bending modes. Therefore, configurations with either low tail volume and high aspect ratio or high fineness ratio can exhibit BFF.

Figure 4 presents a short list of vehicles and vehicle concepts that exhibited some sort of Body Freedom Flutter or were predicted to. In many cases the issue was addressed by limits on the flight envelope. In the Lockheed Martin Polecat, the Body Freedom Flutter was not unstable however it did require that the flight control laws be design to account for the dynamics of the flexible modes of the aircraft.

Low Tail Volume	High Fineness Ratio
 SensorCraft	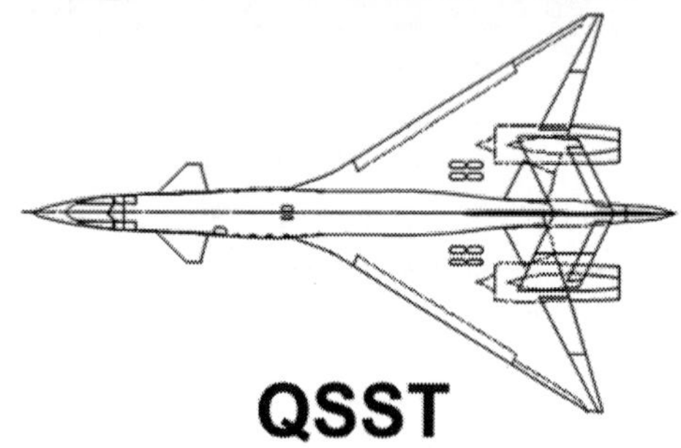 QSST
DarkStar	SR-71A
SensorCraft	Quiet SuperSonic Transport
Polecat	Oblique Flying Wing

Figure 4; List of high aspect ratio and high fineness ratio aircraft that have had Body Freedom Flutter

Research on suppression and exploitation of aeroelasticity and aeroservoelasticity has been ongoing at least since the early 1970's. Figure 5 shows a number of the research and operational vehicles that have been a part of these efforts. In some cases the systems were designed to extend life through load alleviation. Some efforts have focused on opening up the aircrafts operation limits by pushing flutter speeds higher. In other cases, programs had the goal to exploit structural bending to increase maneuver performance.

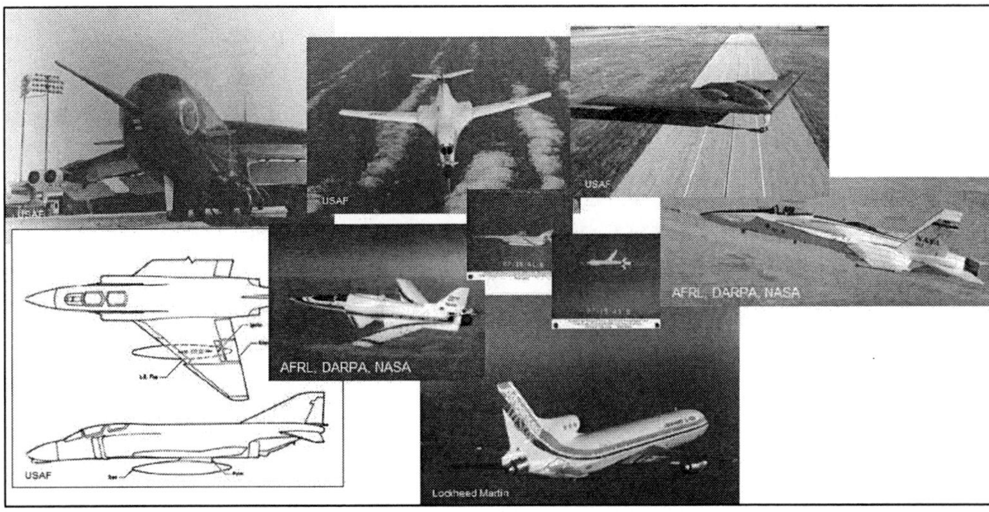

Figure 5; Some of the vehicles that have played a part in Aeroelastic research

The Air Force Research labs performed flutter suppression on the aileron on the B-52 of the Control Configured Vehicle experiment aircraft in 1972.[7] The U.S. Air Force and German Air Force used a modified F-4F that had a movable mass in the external fuel tanks to investigate the effect of mass distribution on flutter in1974.[8]

In the late 1970s the Rockwell/Boeing B-1 was equipped with a Structural Mode Control System that was designed to reduce the loads on the extended forward fuselage. [9][10] The system lowered the loads and improved ride quality for the crew. Lockheed developed a Gust Load Alleviation (GLA) system for the L-1011 to lower the loads on the outboard portion of the wing to extend the life of the structure. [11][12]

The DARPA/AFRL/NASA X-29 with its forward swept planform was designed to exploit wing bending and torsion to enhance maneuverability. Interestingly, forward swept wing configurations also have a tendency for body freedom flutter. In the late 1970s and early 1980s the NASA DAST project performed initial active flutter suppression work using an unmanned target Firebee drone. However the program only had limited success.

In some mission segments the Northrop Grumman B-2 exhibited an oscillatory mode that was categorized as a Residual Pitch Oscillation. The B-2s RPO has been attributed to the oscillation of the shock wave at transonic speed coupling with the rigid body and structural dynamics. [14][15][16]

In 2005 AFRL and NASA modified an F-18 to investigate the ability to exploit the torsional bending of the aircrafts wing to increase maneuverability. The aircraft was re-designated the X-53 Active Aeroelastic Wing. The leading edge flaps were modified such that they could be deflected leading edge up as well as down. The aeroelastic torsion on the wing was used to increase roll maneuverability. [17]

Based on this previous work and the results of the Lockheed Martin BFF IRAD and CRAD programs, the Multi-utility Aeroelastic Demonstrator (MAD) program was initiated.

X-56A

The Multi-utility Aeroelastic Demonstrator (MAD) program had three major objectives: develop a Multi-Utility Technology Test-bed (MUTT) vehicle for active aeroelastic control and Gust Load Alleviation (GLA) Research, develop the MUTT vehicle that will be capable of a variety of configurations, and develop a Multiple Flutter Mode Control System for an unstable vehicle.

The MUTT X-56A is an uninhabited vehicle of 28 foot span with an aspect ratio of 14, a maximum Gross Take Off Weight of 480 pounds, and designed to fly up to 150 knots and 10,000 feet. The program has developed two center-bodies, one stiff set of wings, three flexible sets of wings, a transportable Ground Control Station that houses the simulation, and a transport trailer. Figure 6 presents a three view drawing of the X-56A vehicle and shown in Figure 7 is the entire research system

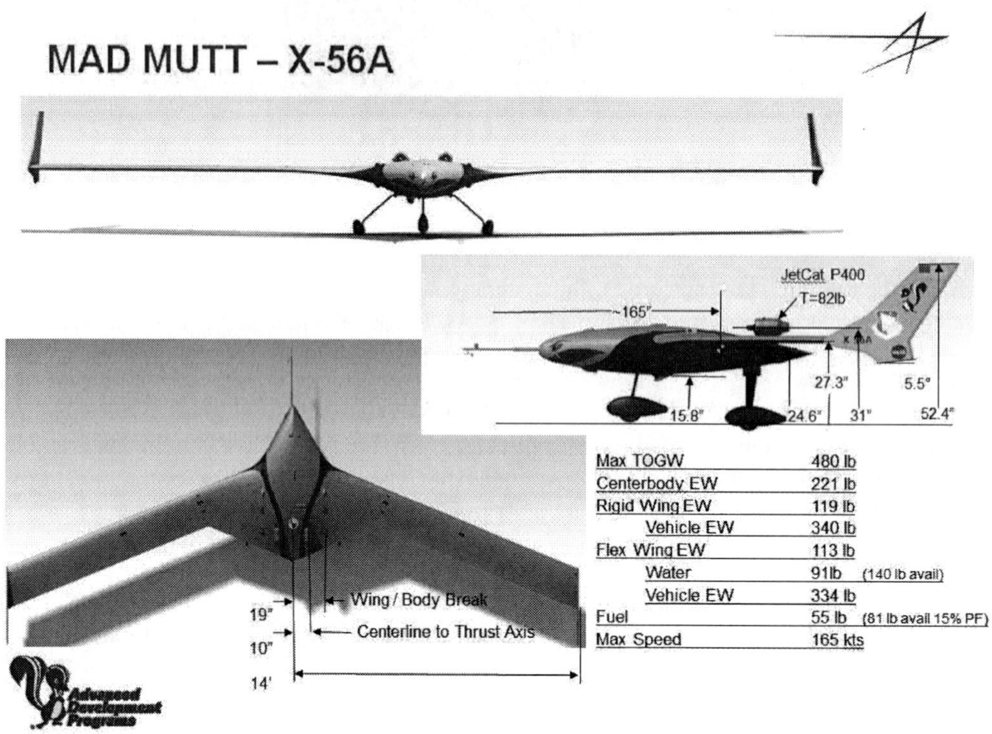

Figure 6; Three View of the X-56A MUTT aircraft

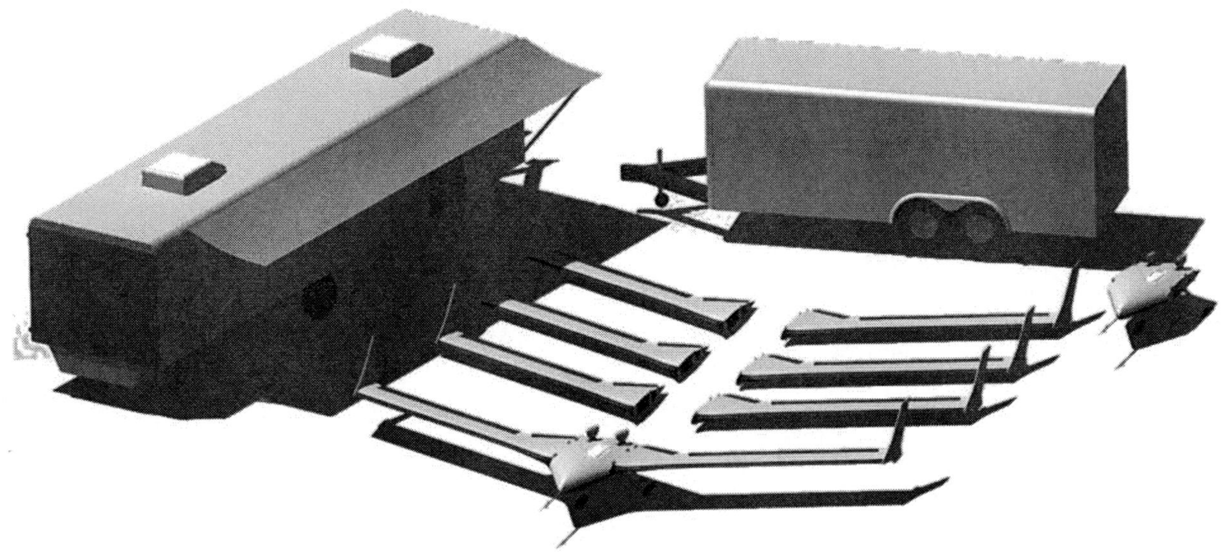

Figure 7; The MUTT was designed to be a complete research system

The X-56A was designed to perform aeroelastic research as well as other high risk flight test. Because of this, the vehicle was designed to maximize future adaptability. The vehicle was designed with all of the avionics housed in the center-body of the aircraft. The center-body was equipped with a BRS-1350 Ballistic Recovery Parachute system to increase the survivability of the avionics. The flight control computer is a "Speedgoat" open architecture COTS computer system. The X-56A is powered by two "Jetcat" P-400 80 pound class turbine engines. The vehicle has an expandable data acquisition system that telemeters data to the vehicles ground station.

Flight operations of the two vehicles known as "FIDO" and Buckeye have taken place at NASA Armstrong Research Center. Shown in Figure 8 are the two aircraft at NASA Armstrong Flight Research Center.

Figure 8; The Two X-56A vehicles "Buckeye" on the left and "FIDO" on the right

The vehicle was designed such that it could accept different wings and support the addition of a tail boom or even a tail arrangement in the future. The landing gear were design such that the main gear could be moved over a range of travel as required by the vehicles configuration. Figure 9 shows the baseline X-56A design and how it might be adapted to perform flight research on a joined wing configured vehicle.

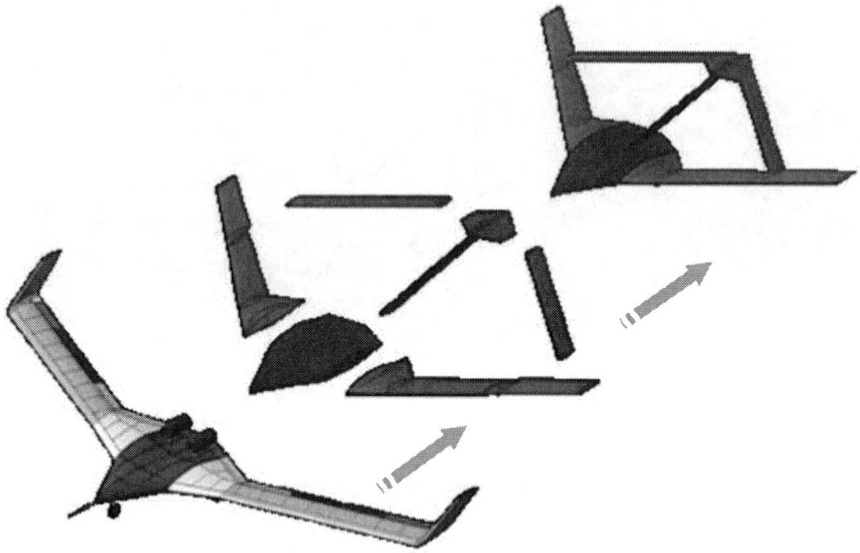

Figure 9; The X-56A was designed to allow it to be flown in multiple configurations

In the case of the X-56A two sets of wings were designed, each to a specific stiffness, with identical internal structure, the only changes were to the material selection of wing skins. The stiff wings could not have flutter within the flight envelope and the flex wings needed to be designed to have three flutter mechanisms with in the flight envelope. The first flutter mechanism required a flutter speed sufficiently above takeoff and landing speed to allow for normal operation and vehicle system checkout.

The normal design process is to determine the strength required through loads analysis and design the vehicle accordingly, then analyze to ensure sufficient aeroelastic stability margin. If the stiffness is not adequate then the design is modified to increase the stiffness (generally over designed for strength) which often increases weight, or the flight envelope is reduced.

The design engineers on the MAD program had to adapt and change their processes to the new mindset of designing for a specific stiffness and assuring adequate strength. Along with the requirement to design for specific stiffness the design needed to account for cost and manufacturability. Other design requirements such as having the vehicle robust and recoverable drove the design, for instance the parachute landing was one of the major strength design cases. Even though it was not easy, in the end the vehicle did end up with a balanced design based on stiffness.

Figure 10 shows the X-56A stiff wing configuration design and demonstration envelopes and the flutter boundary well above the vehicles maximum velocity. This is similar to the conventional design process. Figure 11 presents the same flight envelope with the flexible wings first unstable Rigid Flexible Coupled mode, Body Freedom

Flutter, well within the demonstration envelope. The second and third structural modes are at the limit of the demonstration envelope.

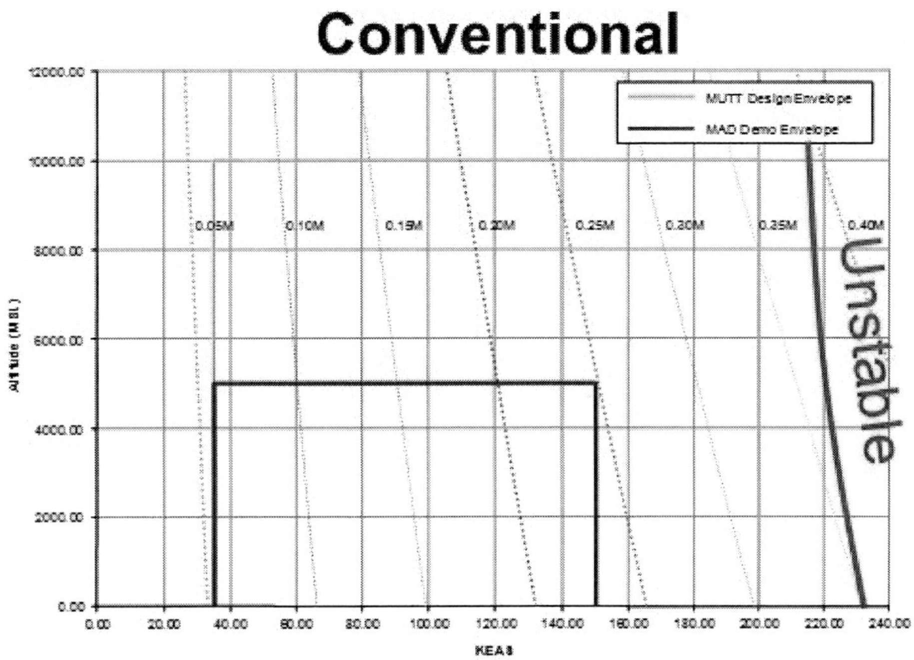

Figure 10; MUTT Stiff wing design with the first flutter mode well outside of the flight envelope

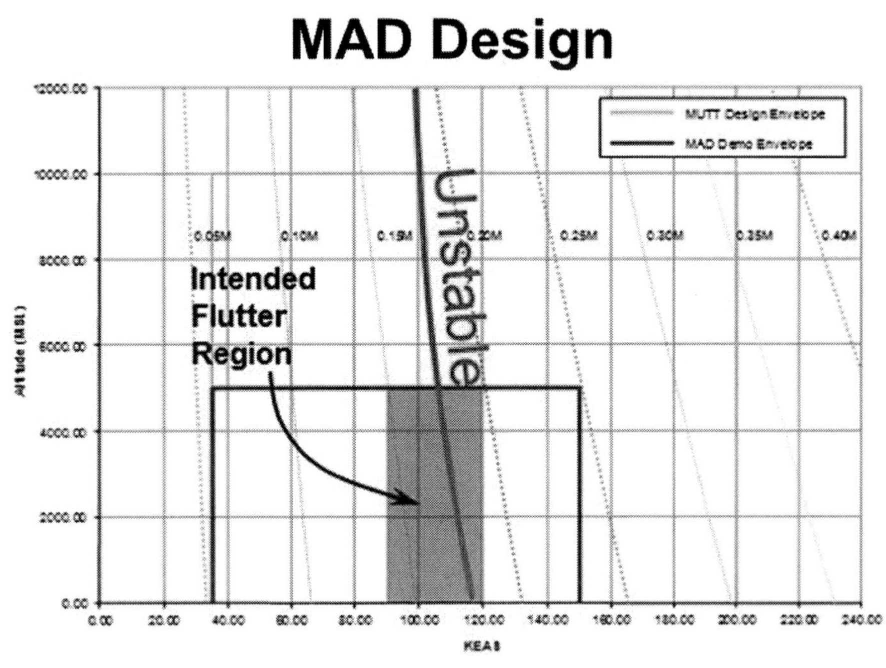

Figure 11; MUTT Flex wing design with the first flutter mode well within the flight demonstration envelope

Lessons Learned

The world of UAVs is very two Dimensional. The fact that the pilot is displaced from the aircraft means that all situational awareness must be transmitted back to the pilot. One consequence of this is that a pilot's vision is limited in field of view and is monocular. Many aviators have been able to overcome the lack of binocular vision. For instance Wiley Post who lost his left eye early in his flying career was able to accommodate the loss of depth perception.

By its nature, first person UAV flying has to have a means to replicate or transmit representative Fields of View (FOV) to a pilot remote from the vehicle. This is required for situational awareness, attitude information, rates and acceleration determination, command feedback status, and decision making. The X-56A was designed with these considerations in mind. The situational awareness of some of the states of the vehicle can easily be lost when flying a UAV even with a good forward view camera and well thought out heads down displays. Two means of situational awareness were utilized; a forward looking camera, and a virtual world view for X -56A as a method of providing redundancy. The camera was chosen such that the FOV mimicked a monocular vision perspective of roughly 17 by 35 degrees FOV to keep from having fisheye or soda straw images displayed to the pilot. Having the FOV close to what a normal human experiences allows for correlation and perception to be easily obtained while increasing situational awareness. The virtual world view was designed with the same perspective as the camera so that quickly transitioning from one to the other did not cause confusion.

Camera mounting in the UAV is worthy of careful consideration. First, the camera should be mounted close to the position that a person would be in a manned aircraft, near the CG or center of rotation. Second, the camera should be bore sited with respect to the fuselage in pitch, roll, and yaw. And third, the camera if possible should have some form of vehicle structure in its field of view, but not so much as to render the camera useless for its intended purpose. The X-56A flight test boom fit the bill perfectly for this third consideration. The virtual world perspective is driven by the navigator system on the X-56A which is bore sited to the airframe much like the camera. Airframe reference was achieve by utilizing fixed display indicators within the virtual world displays; similar to old style Attitude Display Indicator with pointers attached to the instrument case. For camera FOV without any frame of reference the same indicator used in a virtual display will suffice. Following these considerations, a pilot can with practice accurately determine rates, accelerations, and mode shapes about all three axis of an aircraft. The pilot can determine stable versus unstable behaviors; one of the first steps in deciding if pilot input is required.

The sense of velocity along the vehicle flight path can be very difficult to determine depending on perspective (close to the ground or far away from it) when viewed through a camera or virtual world. The X-56A utilizes an energy state cue to assist the pilot in maintaining a proper energy state i.e. airspeed for the vehicle flight regime. The energy dot cue is displayed in terms of flight path angle to allow the pilot better prediction of the trimmed state of the vehicle. Unfortunately the energy rate cue does not give the pilot a sense of the aircrafts higher frequency accelerations in all axes. Through simulation and experience it was found that just having an acceleration cue without some additional automation such as inner loop speed control resulted in a high pilot workload.

The X-56A the Ground Control Station maximized the reuse of displays. The Heads Up Display (HUD) design started with Mil-Std-1787 and the Heads Down Displays (HDD) were based on SAE ARP-4102-7. Figure 12 shows the pilots and Test Directors stations in the MAD Ground Control Station. The HUD display is overlaid on the large forward pilot camera view. The HDD includes a synthetic view of the runway. Not seen in the simulated view in Figure 12 is the flight test air data boom.

Figure 12; MAD Ground Control Station Pilot and Test Director Station

Behind the pilots in Figure 12 on each side are additional displays that include a moving map and other data entry capabilities such as the Flight Test Aid control panel. The two HDD are shown in Figure 13. The primary displays and controls, including the electronic flight displays, engine displays and controls and integrated Caution and Warning indicators are on the forward display on the right in figure 13, the moving map display and flight test aids data entry panel are located on the side display on the left. The moving map could be configured to show airspace boundaries, turn rate indicators, and the loss of communication plan.

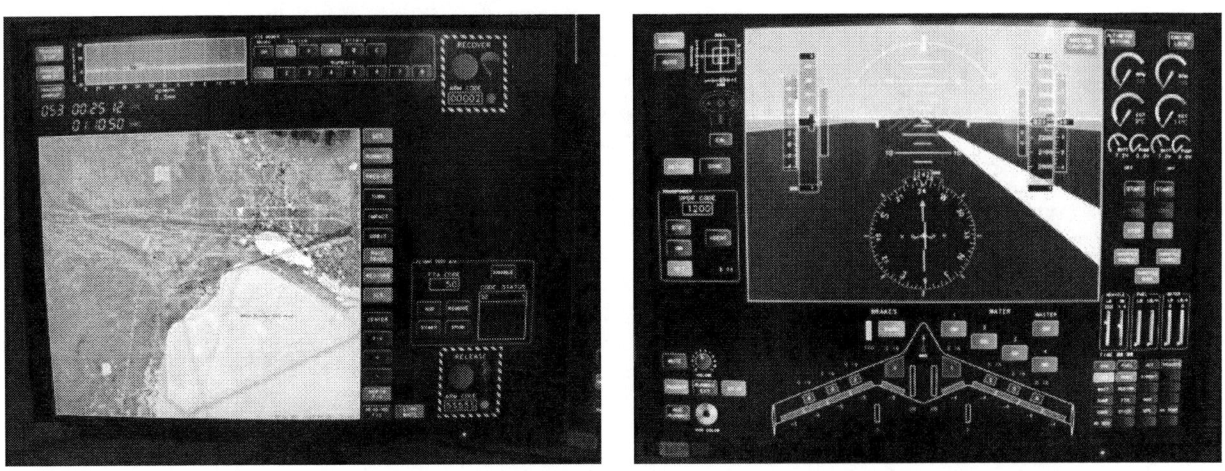

Figure 13; The two pilot Heads Down Displays in the X-56A Ground Control Station

By developing the real-time simulation capability early in the program, pilot feedback could be used to tailor the controls and displays. The displays were developed from the standards and modified where necessary to address pilot feedback. Failure Mode Cause and Effects Analysis and systems analysis also guided the design. A command and achieved feedback philosophy was used. The commands displayed to the pilot were sent from the aircraft to the pilot station to ensure the pilot understood what the aircraft was trying to achieve. This also allowed for orderly transfer of control of the vehicle from the Pilot in Command to the Test Director if required.

Figure 14 shows the Electronic Flight Instruments displayed to the pilot on the heads down display. Figure 15 is an enlargement of the center of the display to show the commanded flight path (the square corners) and the achieved flight path marker.

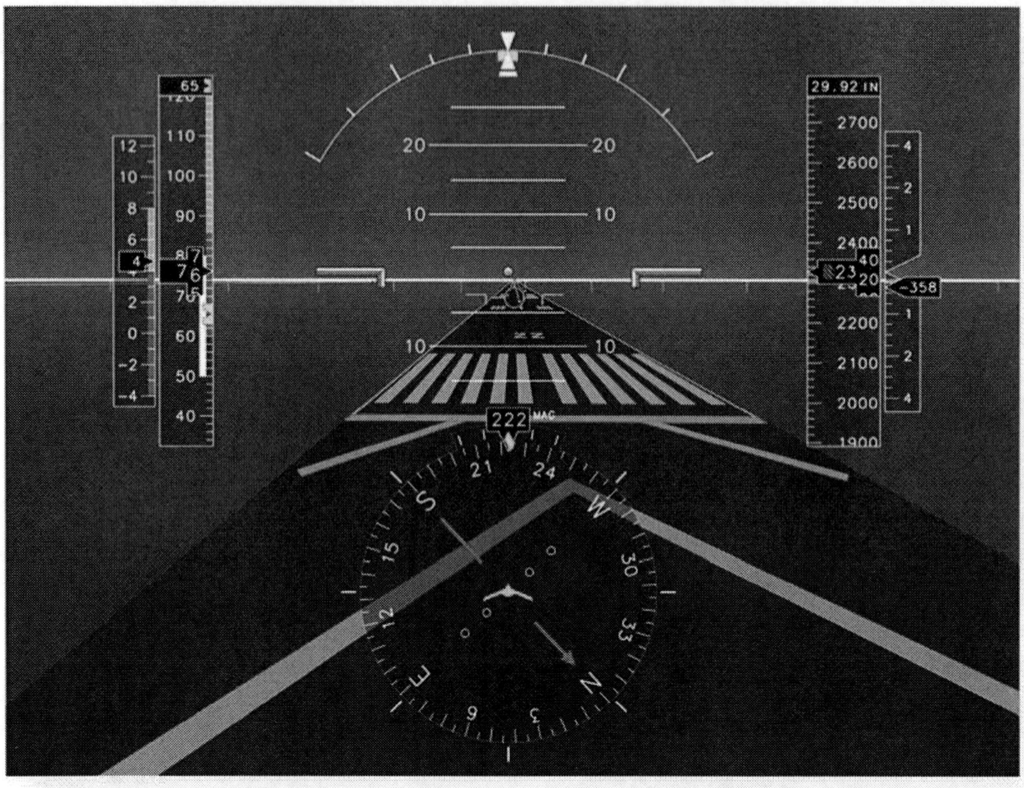

Figure 14; Heads Down pilot display with synthetic view

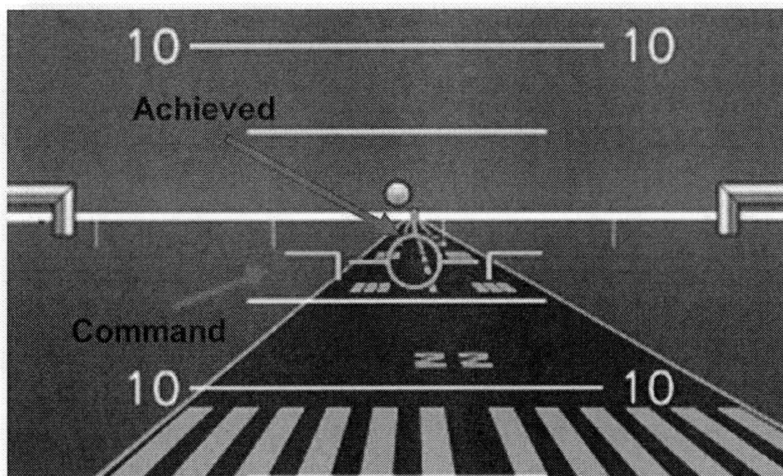

Figure 15; Expanded view of display showing Command and Achieved values for flight path

One additional part of the X-56A situational awareness suite was the addition of an on-board microphone. While it may not be intuitive that a microphone would assist situational awareness, it does in several ways. The audio feedback was very compelling for the pilots to know if the vehicle was on the ground or in the air. On the ground it gave good feedback for engine RPM. During the pre-engine start power up of the vehicle, the crew chief would talk to the pilot and test conductor through the aircrafts microphone instead of the radio. This worked very well, however it caused great confusion and concern for at least one photographer when the crew chief started talking to the aircraft.

Figure 16; Crew Chief and X-56A during engine start

The challenge for any pilot in the loop UAV is command and control feedback implementation. This has to be designed in concert with vehicle displays and FOV considerations. Command path and feedback requirements are driven by latency, vehicle purpose, vehicle complexity, failure mode accommodation, and pilot perspective (first person or external). The X-56A utilizes a gamma command, bank angle command, and yaw rate command implementation with either speed command or throttle command for power management. The advantages are; less upfront autonomous vehicle programming is required, real time pilot inputs can be made without software changes, and potential system errors can be recoverable.

Some may challenge the definition of pilot in the loop with respect to UAVs. However, based on vehicle state information; when pilot input or lack of pilot input, or misapplied pilot input; cause desired or undesired actions, which in turn the pilot further reacts to, the pilot is in the loop. Based on X-56A control law inner loops maintaining stability, the pilot is commanding outer loop parameters. The most important aspect of the outer loop is maintaining the appropriate energy state for each flight condition.

In some cases a higher level augmentation may need to be turned off during specific tests. For instance, based on simulation rehearsals of testing near the flutter boundaries it was determined that the use of speed mode was not appropriate due to the inability to slow down the airplane while waiting for the speed command to roll off. Direct engine RPM command (PLA) was necessary as the control loop and integrator in speed command loop would not allow for a rapid reduction in speed out of the flutter region.

Review of the X-56A Flight 1 landing revealed approximately a 2 Hz nose bounce upon nose wheel touchdown that was readily apparent to the pilot. Display fidelity was great enough that stability was quickly ascertained, however pilot input was required to ensure the system did not become unstable. Due to the X-56A having a gamma command system, direct pilot control of pitch rate is unobtainable. Fading out of the pitch axis inner control loops for on ground operations was designed to be in the 40-45 KEAS range, below the alpha limiter speed for all aircraft weights. The resulting decision was to either reduce or add energy to the vehicle, i.e. perform a go around or slow as quickly as possible to 40-45 KEAS, a speed range experienced on many previous taxi tests. Results are shown in figures 16 and 17. The pilot made inputs to reduce the energy state of the vehicle which resulted in the dampening of the nose bounce ground interaction of the aircraft. Of note, the pilot did not try and modulate brakes or gamma command to dampen the pitch rate, as even if they could, the frequency was too high, and a resultant out of phase input would most likely have occurred with undesired results. Understanding the control system you are operating, the limitations, and the extent of its authority, along with the feedback provided are key factors in determining when the pilot should get in the loop with an UAV.

There has been a great deal of debate in the industry and academia regarding the issue of cockpit automation and its effect on piloting. [(18)] Does automation reduce a pilots skills or does it change the pilot's decision making process? UAV pilots may also be more apt to abstain from making inputs versus their manned aircraft counterparts when pilot inputs are called for. This could be due to potential perception of detachment from UAV, lack of kinesthetic cues, lack of risk of bodily harm, information overload, channelized attention, or non-traditional control inputs lend to a greater number of incidents that a UAV could have been saved, but for the lack of pilot interaction. If one assumes the information overload and channelized attention are common for manned and unmanned aircraft, than what is left is the perception of detachment, lack of kinesthetic cues, and lack of risk of bodily harm. These possibly contribute to UAV pilots watching as their aircraft continue toward trouble versus active pilot engagement. Further study is warranted to ascertain objectively the UAV pilot perception of when to get in the loop in what is a challenging aircraft flying environment.

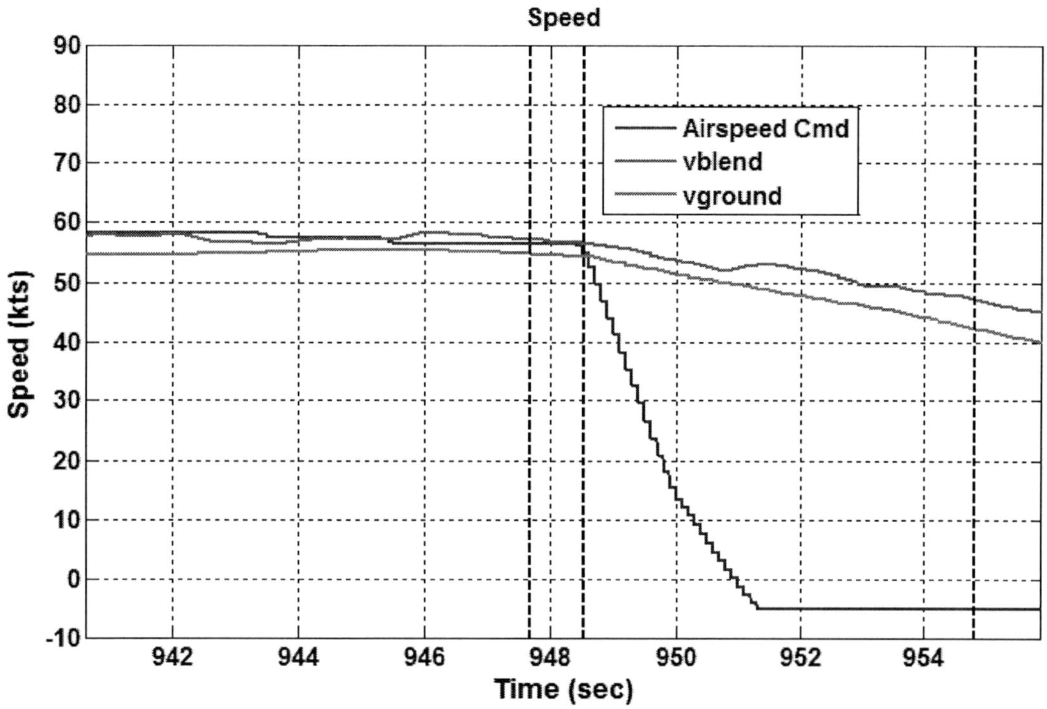

Figure 16; Speed command and achieved data from the landing of the first flight of the X-56A

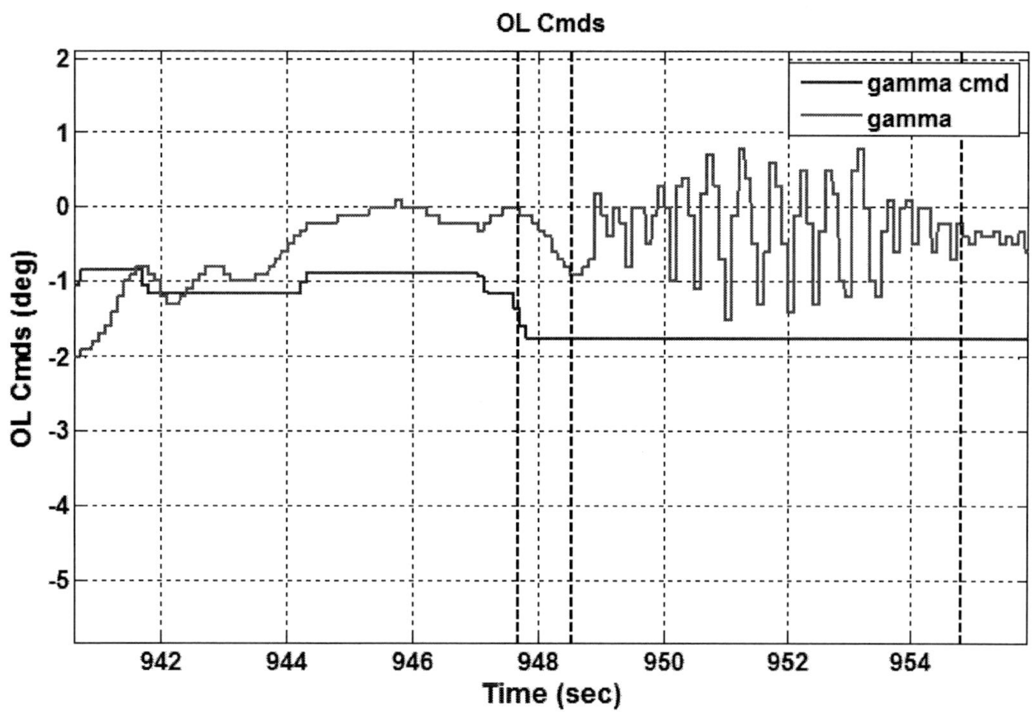

Figure 17; Gamma command and achieved data from the landing of the first flight of the X-56A

The Path forward

As we look to improve performance using Fly-By-Wire technology to reduced static stability, and Active Flutter Suppression to reduce an aircraft's structural weight, we need to understand and tailor the current rules of airworthiness.

We will need to be able to test the Active Flutter Suppression system for certification. This will require the development of higher fidelity N-DOF real time simulation capabilities that can simulate the modal vibration of the aircraft and the stimulation of the structural sensors. Our simulations will need to be able to provide higher fidelity models of system failures as well.

We need to determine what level of design margin we should design to when it comes to structural mode control, the 8 dB and 60 degrees of phase for structures or the 6 dB and 45 degrees of phase for flight controls.

Just as we have developed Built-In-Tests and health monitoring system to ensure that a Fly-By-Wire aircraft will function correctly, we must develop methods to test that the aircraft's on board structural model matches the aircraft in its current configuration. We may be able to identify a sensor that is not operating as expected but how do we know if the structural dynamics have changed? Perhaps a pre-flight Ground Vibration Test will be required.

Recently the FAA issued an Airworthiness Directive to a commercial aircraft. Specifically stated a risk of "divergent flutter during a high g-load maneuver in combination with certain system failures." So these questions are not something that we should be addressing in the long term, we need to be working on them now.

Once we have convinced ourselves we have the correct requirements for the design and we have built in methods to test the systems on the ground, we must look at our flight test plan to best understand how to test the vehicle in the air. We now must ask ourselves what is the test process to be used and what is the build-up approach. Is the conventional test build up approach used for manned aircraft the correct or at least the safest testing method?

What other alternatives should we look at? Should we examine the use of optionally piloted and remotely piloted vehicles to open the envelope? For the X-56A, we design "Stiff" and "Flexible" wings to allow for system checkouts to be done without the additional issues of flutter. So do we design our new vehicles with a structure with removable stiffeners?

For years flutter engineers have been trying to create some sort of "Flutterometer" without success. It is one thing to research flutter in a test environment with engineers on the ground monitoring real time telemetry, Power Spectral Density displays, and utilizing methods of damping extraction. These engineers are trained to look for the first indication of a mode coupling and it's tell tail sinusoidal wave form on accelerometers as damping is lost.

Figure 18 shows a notional Power Spectral Density (PSD) plot and signal decay plot similar to what an aeroelastics engineer would be looking at during developmental flight test. This type of information does not lend itself to display in the cockpit; however due to the explosive nature of flutter a pilot on board the aircraft needs to know the state of the vehicle. With the development of Active Flutter Suppression and the existence of on-board structural models, it may now be possible to develop methods to give the pilot a measure of the vehicles flutter margin. But what does this look like? We as a community need to decide. With flutters rapid onset and catastrophic behavior it does not make sense to think of adding a "Flutter" light! Figure 19 shows how not to communicate a flutter issue to a pilot.

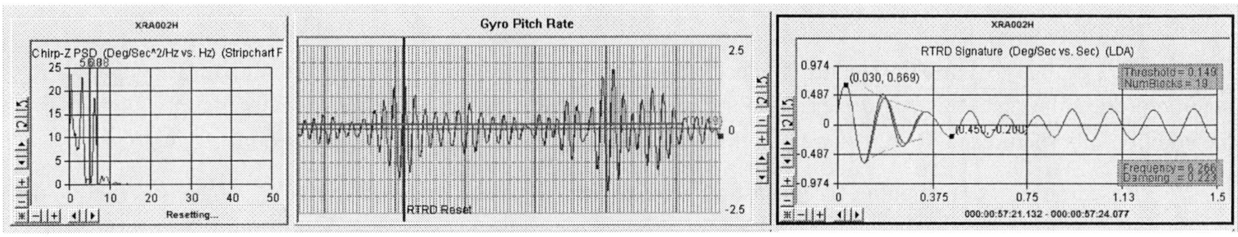

Figure 18; Notional Power Spectral Density plots used in flight test

Figure 19; Due to the nature of flutter we cannot think of it in the same way as other faults

The flutter community has been trying for decades to develop a technology to measure and analyze flutter margin in flight. However, the knowledge used in the development of these type of suppression systems may allow us to better inform the pilot of the loss of damping of a structural mode. Can on board analysis be used to derive a measure of the stability of the controlled system? Figure 20 shows one possible way to show the state and the trend for an operational vehicle.

Figure 20: A notional "Flutterometer" showing damping margins at multiple points on the vehicle

Conclusion

One of today's limitation to aircraft performance is the result of aerodynamic and structural dynamic coupling. The solution to this Rigid Flex Coupling has been to increase the structural stiffness of the vehicle through the addition of structure and therefore weight, the other alternative has been to limit the aircraft's envelope.

Research has been ongoing for the last 50 years to find solutions to these limitations. Currently the X-56A is being used by Lockheed Martin under contract to AFRL and supported by NASA Armstrong Flight Research Center to further this research. The X-56A is an unmanned aircraft which allows higher risk flight research to be conducted.

The operations of the X-56A have pointed to some lessons learned with respect to the operation of a research UAV. These lessons include pilot situational awareness, the pilot command and control types, and issues in assessing when and how much a pilot should get into the control loop.

Along with the lessons learned from the operation of the X-56A there are also questions that need to be addressed for the long term application of the advancing field of active flutter suppression. We as a community of pilots and engineers need to define the future requirements for airworthiness. We must define the best and safest methods to test the aircraft that employ this technology. And we must provide this insight to the regulatory agencies that regulate the vehicles operations,

Acknowledgements

Support for this program was provided by Mr. Pete Flick and the Air Force Research Laboratory; additionally the authors would like to thank NASA Armstrong Flight Research Center for their efforts in supporting the flight test of the X-56A MAD MUTT.

References

[1] Nicolai, L.M., Hunten, K., Zink, S., Flick, P., "System Benefits of Active Flutter Suppression for a SensorCraft-Type Vehicle," AIAA/ISSMO Multidisciplinary Analysis Optimization Conference, 13 - 15 September 2010, Fort Worth, Texas AIAA-2010-9349

[2] Beranek, J., Nicolai, L., Buonanno, M., Burnett, E. L., Atkinson, C., Holm-Hansen, B., Flick, P., "Conceptual Design of a Multi-utility Aeroelastic Demonstrator" AIAA/ISSMO Multidisciplinary Analysis Optimization Conference, 13 - 15 September 2010, Fort Worth, Texas AIAA-2010-9350

[3] Love, M., Zink, P., Wieselmann, P., Youngren, H., "Body Freedom Flutter of High Aspect Ratio Flying Wings," AIAA-2005-1947

[4] Burnett, E.L., Atkinson, C., Beranek, J., Sibbitt, B., Holm-Hansen, B.T., Nicolai, L., "NDOF Simulation Model for Flight Control Development with Flight Test Correlation," AIAA Modeling and Simulation Technologies Conference, 2 – 5 August 2010, Toronto, Canada AIAA-2010-7780

[5] Holm-Hansen, B.T., Atkinson, C., Beranek, J., Burnett, E.L.., Nicolai, L., Youssef, H., "Envelope Expansion of a Flexible Flying Wing by Active Flutter Suppression," Association for Unmanned Vehicle Systems International Conference, August 2010,

[6] Tilmann, C.P., Flick, P.M., Martin, C.A., Love, M.H., "High-Altitude Long Endurance Technologies for SensorCraft," RTO Paper MP-104-P-26, RTO AVT-099 Symposium on Novel and Emerging Vehicle and Vehicle Technology Concepts, 7 – 11 April 2003, Brussels, Belgium.

[7] Roger, Kenneth L., Hodges, Garold E., Felt, Larry, "Active Flutter Suppression – A Flight Test Demonstration" *Journal of Aircraft*, Vol. 12, No. 6, 1975 PP. 551-556

[8] Sensburg, O., Honlinger, H., Noll, T.E., Huttsell, L. J., "Active Flutter Suppression on an F-4F Aircraft" *Journal of Aircraft*, Vol. 19, No. 5, 1982 PP. 354-359

[9] O'Connell, R.F., Messina, A.F., "Development of an Active Flutter Margin Augmentation System for a Commercial Transport", Journal of Guidance and Control, Vol. 3, No. 4, 1980, PP 352-360

[10] Wykes, J.H., Mori, A.S., Borland, C.J., "B-1 Structural Mode Control System," AIAA Aircraft Design, Flight Test, and Operations Meeting, 7-9 August, 1972, Los Angeles, California, AIAA 72-772

[11] Guinn, Wiley A., Rising, Jerry J. Davis, Walt J., "Development of an Advanced Pitch Active Control System for a Wide Body Jet Aircraft," NASA Contractor Report 172277, 1 February 1984

[12] O'Connell, R.F., "Design, Development and Implementation of an Active Control System for Load Alleviation for a Commercial Transport Airplane," 49th AGARD Structures and Materials Panel, 9 October 1979, Cologne, Germany

[13] Wykes, John H., Byar, Thomas R., MacMiller, Cary J., Greek, David C., "Analyses and Tests of the B-1 Aircraft Structural Mode Control System," NASA Contractor Report 144887, January 1980

[14] Jacobson, S. B., Britt, R. T., Freim, D., Kelly, P., "Residual pitch oscillation (RPO) flight test and analysis on the B-2 bomber", 39th AIAA/ASME/ASCE/AHS/ASC Structures, Structural Dynamics, and Materials Conference and Exhibit, 1998

[15] Britt, R. T., Volk, J. A., Dreim, D. R., Applewhite, K. A., "Aeroelastic Characteristics of the B-2 Bomber and implications for Future Large Aircraft.," RTO AVT, Ottawa, Canada 18-20 October 1999, RTO MP-36

[16] Britt, R. T., Jacobson, S. B., Arthurs, T. D., "Aeroservoelastic Analysis of the B-2 Bomber Journal of Aircraft", 2000, Vol.37: 745-752

[17] Pendlton, E., Flick, P., Paul, D., Voracek, D., Reichenbach, E., Griffin, K., "The X-53 A Summary of the Active Aeroelastic Wing Flight Research Program," AIAA Structures, Structural Dynamics and Materials Conference, 23-26 April 2007, Honolulu, Hawaii, AIAA 2007-1855

[18] Jacobson, S. R., "Aircraft Loss of Control Causal Factors and Mitigation Challenges", AIAA Guidance, Navigation and Control Conference, Toronto, Ontario, Canada, 2010, AIAA 2010-8007

F-22 Line-in-the-Sky Automatic Ground Collision Avoidance System

(LIS AGCAS)

Steven Rainey, Lockheed Martin, Edwards AFB, CA
Major David Allamandola, United States Air Force Test Center, Edwards AFB, CA
Mr. Charles Webb, United States Air Force Test Center, Edwards AFB, CA

ABSTRACT

This paper provides an overview of an Automatic Ground Collision Avoidance System (AGCAS) as implemented in the Lockheed Martin F-22. This paper will discuss the design philosophy, test planning, test execution, test results to include pilot-vehicle-interface issues, fielding concerns, and recommendations. The AGCAS as implemented on the F-22 was developed in an attempt eliminate controlled flight into terrain (CFIT) mishaps in the F-22 fleet. Testing was divided into two different phases with a developmental phase that only tested the performance of the recovery algorithm and an operational phase that allowed more data to be collected with respect to how the system integrated with the normal operations of a tactical fighter. The implementation of the AGCAS makes proper utilization of the system difficult and in some cases the system must be turned off to complete mission essential tasks. However, testing showed that the recovery performance of the F-22 AGCAS algorithm when set and utilized properly is excellent. Ultimately with extensive pilot training to overcome the difficulties in utilizing the system, the F-22 LIS AGCAS will save lives and prevent CFIT accidents in the F-22 fleet.

ACRONYMS, ABBREVIATIONS, SYMBOLS

AGCAS – Automatic Ground Collision Avoidance
AoA – Angle of Attack
ATLAS – Aircraft Trim Linearization and Simulation
BME – Benign Maneuvering Envelope
CFIT – Controlled Flight into Terrain
COTS – Commercial Off The Shelf
CSAF – Chief of Staff of the Air Force
DTED- Digital Terrain Elevation Database
EMD – Engineering Manufacturing Design
FAB – Function Access Button
FLCS – Flight Control System
GINS - Global Positioning System/Inertial Navigation System
HAE – Height Above Ellipsoid
HUD – Head Up Display
ICAW – Indications Cautions Advisories and Warnings
ICP – Integrated Control Processor
IMC – Instrument Meteorological Conditions
IVSC – Integrated Vehicle Subsystem Controller
LAW – Low Altitude Warning
LIS – Line in the Sky
LUFD – Left Upfront Display
MK – Menu Key
OFP – Operational Flight Program
OSD – Office of Secretary of Defense
PANT – Post Abort Nonresponsive Test
PVI – Pilot Vehicle Interface
RTB – Return To Base
SWIM – System Wide Integrity Monitor
TTFU – Time to Fly-Up
TSM – Time Safety margin
VFR – Visual Flight Rules
VMS – Vehicle Management System

INTRODUCTION

The baseline design philosophy of any AGCAS originated from NASA and F-16 AGCAS development, which has been ongoing in various stages since 1994. Some of the design considerations have evolved over the years, but the basic design philosophy remains the same and has been considered universal to the F-16, F-22, and F-35 aircraft (reference 1). The primary guiding principles, in priority order, have been:

1. Do no harm to the pilot or aircraft.

2. Do not interfere with normal operations.

3. Prevent ground collisions.

The goal for every AGCAS design was to prevent controlled flight into terrain (CFIT) mishaps by providing an automated aircraft recovery mechanism to protect the pilot and airframe in situations where the pilot was task-saturated, disoriented, or incapacitated. The system must provide this capability without interfering with the pilot's ability to complete the assigned training or combat mission. This equated to minimizing nuisance recovery maneuvers while retaining freedom for the basic flying techniques. The pilot in command was ultimately responsible for safety and the system was not intended to provide 100 percent protection. Additionally, the pilot must be able to disengage the AGCAS or interrupt a nuisance recovery.

The AGCAS was proposed for implementation on the advanced tactical fighter since the early 1990's and was intended to be based on a digital terrain elevation database (DTED). At the time of the F-22 LIS AGCAS test effort, the F-22 System Program Office (SPO) directed that any AGCAS system on the F-22 would be a flight critical system. There was insufficient throughput available in any of the flight-critical safety redundant systems on the aircraft, such as the flight control system (FLCS), or the integrated vehicle subsystem controller (IVSC), etc., to support a dual-redundant DTED-based AGCAS. The system implemented on the F-22 was instead based on a LIS, which was an altitude level manually selected by the pilot. The LIS AGCAS would perform an automatic fly-up to avoid flying below the LIS. The LIS AGCAS required less computational resources and therefore was compatible with existing hardware.

The LIS AGCAS was developed to meet OSD/CSAF guidance to field a system that reduced F-22 controlled flight into terrain (CFIT) mishaps. This system was new flight control software integrated with existing hardware. It was designed to initiate and execute a recovery maneuver using longitudinal and lateral stick commands to bring the aircraft to an upright attitude and prevent descent below a pilot-selected altitude floor.

A DTED solution was desired over a LIS solution because a DTED-based AGCAS required less pilot interaction and offered less nuisance/mishap potential. A DTED-based system was constantly aware of the aircraft's position relative to the ground. The LIS solution of the AGCAS relied on regular pilot interaction to select a level that was appropriate for the terrain in the local area of operations. The main limitation of a LIS solution is the excessive required pilot interaction. The pilot must set the line, remember where the line was set, and then remain cognizant of the aircraft's position relative to the line while focused on other demanding mission tasks. Pilots are unable to accomplish this 100 percent of the time due to human limitations as demonstrated by flight test. There is some question as to whether this is acceptable due to unintended safety considerations (e.g. midair collision, loss of situational awareness during IMC fly-ups, or peacetime training scenarios in congested airspace).

The LIS AGCAS replaced the low altitude warning (LAW) functionality of the F-22 aircraft. The legacy LAW functionality triggered an "*Altitude, Altitude*" tactical alert when the aircraft descended through a barometric altitude floor manually selected by the pilot. This legacy LAW functionality was commonly used during tactical phases of flight where the LAW was set at a tactically significant altitude (e.g. fight floor, transition altitude, minimum safe altitude, etc.). Pilots were alerted when they descended through the LAW setting. The manually-selected altitude level for LIS AGCAS was set using the same procedures for selecting a LAW setting with the legacy LAW functionality. The legacy function is no longer available with LIS AGCAS.

In this paper, the LIS AGCAS altitude level will be referred to as "AGCAS line," even though the aircraft still displayed the "LAW" terminology next to the numerical value shown in the integrated control panel (ICP) and head-up display (HUD). See the next section for more detailed information on selecting the AGCAS line. The terms "recovery" or "fly-up" are used to define when LIS AGCAS assumed control of the aircraft to prevent a penetration of the AGCAS line. These terms are not to be confused with the actual RECOVERY mode or the FLY-UP phase of the RECOVERY mode, which refer to more specific and detailed LIS AGCAS actions

TEST ITEM DESCRIPTION

The LIS AGCAS was designed to reduce the risk of CFIT mishaps. The LIS AGCAS was a software change only and was integrated with existing hardware. The software performing the basic LIS AGCAS function resides within the vehicle management system (VMS). To minimize software changes, all LIS AGCAS functions reside in the FLCS and IVSC computer software. The LIS AGCAS was designed to operate as an outer loop function of the F-22 flight controls. Pitch and roll commands were introduced in the stick paths, which were interpreted by the FLCS as a pilot input. Therefore, the existing F-22 inner-loop control laws were unaffected and all automated maneuvering limits were still in place.

The overall LIS AGCAS capability is a collection of three main subroutines within the F-22 flight control system architecture. The main functions that will be discussed in this paper will be the Predicted Recovery Altitude, computation of the Time-To-Fly-Up (TTFU) parameter, and actual aircraft recovery performance. Finally, a brief description of the visual and audio cues available to the pilot will be presented along with the different operating modes of LIS AGCAS.

Recovery Altitude Prediction

The Predicted Recovery Altitude function determined what the minimum altitude would be if LIS AGCAS was to initiate a fly-up at that particular instant. This function resided in the FLCS and was computed at 16 Hertz via look-up tables that were generated using a high fidelity F-22 6-Degree of Freedom aircraft trim, linearization, and simulation (ATLAS) performance model, which was validated during EMD testing. These tables were used to determine a recovery altitude at five discrete altitudes (0 feet, 5,000 feet, 10,000 feet, 15,000 feet, and 20,000 feet) with interpolation for altitudes between these tables and extrapolation for settings greater than 20,000 feet.

The recovery altitude function was based on steady state, 1 g maneuvers where the inputs were aircraft dive and bank angle, Mach number, and Global Positioning System/Inertial Navigation System (GINS) Height Above Ellipsoid (HAE) altitude. These tables were set at evenly spaced intervals with additional interpolation points set where there was inherently a greater amount of model uncertainty (transonic Mach numbers, near vertical dive angles). The recovery altitude function did not account for additional factors such as elevated g and off-design gross weight. Adjustments for these factors were incorporated in the TTFU function.

Time-To-Fly-Up Calculation

The TTFU was a vital piece of LIS AGCAS. The recovery altitude was computed and then compared to the AGCAS line to determine a TTFU. The TTFU was defined as the amount of time that the aircraft could remain at the current state before penetrating the AGCAS line using the LIS AGCAS recovery maneuver. The TTFU was calculated at 64 Hertz and accounted for off-design gross weight and elevated g up to the full g capability of the airplane.

The TTFU was the metric used for triggering visual and aural cues that were presented to the pilot if it was determined that a fly-up was imminent. The ALERT phase began when TTFU was 4 seconds with a "*Pull Up, Pull Up*" tactical alert. The dangerous condition tone immediately followed the "*Pull Up, Pull Up*" until the fly-up initiated or until the pilot took corrective action. If no corrective action was taken, the recovery maneuver would begin when TTFU reached 0 seconds.

Dynamic maneuvering may shorten the timeline such that the ALERT phase would not last the entire 4 seconds. For instance, if a pilot had a LAW of 10,000 ft set and was flying level at 11,000 ft, the TTFU is essentially infinite; however, if he were to roll inverted and pull max g, TTFU rapidly becomes less than 4 seconds.

To address concerns about unintended fly-ups while in close formation or close proximity to other aircraft such as during a battle damage check, the Benign Maneuvering Envelope (BME) was developed. The BME was defined as dive angle less than 10 degrees, and bank angle less than 60 degrees. In BME, TTFU is 6 seconds instead of 4 seconds. The 6 second ALERT phase was intended to give sufficient time for the pilot to recognize an impending fly-up and prevent a fly-up that was potentially unsafe (i.e., occurring while flying in close formation) or that was otherwise unnecessary. In the BME when TTFU equals 6 seconds, the ALERT phase is linear and the fly-up will not begin until 6 seconds has elapsed even if dynamic maneuvering resulted in penetrating the LIS AGCAS line. This afforded the pilot enough time to avoid a potentially dangerous situation by flying away or turning the system off. This mechanization could potentially result in a penetration of the AGCAS line under certain conditions, but allows for a pilot to prevent a potentially unnecessary fly-up.

Recovery Maneuver

If a penetration of the AGCAS line was predicted, LIS AGCAS initiated an automated recovery using longitudinal and lateral stick command paths. During a recovery, the *fly-up* phase of the RECOVERY mode of LIS AGCAS attained a positive rate of climb, where the system transitioned to the *fly-out* phase of the RECOVERY mode using autopilot commands to attain a 30-degree-banked climb to 5,000 feet HAE over the AGCAS line while maintaining a speed of approximately 300 KCAS (±25 knots). Upon reaching the target altitude the aircraft would loiter in a steady, 30-degree-banked turn holding altitude and airspeed until the fly-up was terminated by the pilot. The fly-up could be terminated by the pilot at any point by depressing the paddle switch described below.

If the aircraft was not in an upright, wings-level attitude at the initiation of the *fly-up* phase, LIS AGCAS commanded a roll with a roll rate as high as 150 degrees per second. The command was ramped up at a rate of 480 degrees per second squared. If the recovery initiated when the actual load factor was above 80 percent of the maximum allowable load factor of the aircraft, then LIS AGCAS commanded a maximum roll rate of 25 degrees per second until the load factor had decreased below 80 percent of the maximum allowable load factor.

The amount of load factor commanded by LIS AGCAS during the *fly-up* phase was a function of bank angle and "g-feedback". If a roll command was required from an inverted attitude, to return the aircraft to wings level during the *fly-up* phase, LIS AGCAS only commanded 1 g during the roll maneuver and ramped up toward 4 g when the aircraft reached 135 degrees of bank. Once the aircraft reached 90 degrees of bank, LIS AGCAS commanded the nominal 5 g to return the aircraft to a positive rate of climb while continuing to roll wings level. This nominal 5 g command could be augmented by the g-feedback signal if LIS AGCAS determined that additional performance was required to return the aircraft to a positive rate of climb before penetrating the AGCAS line. The g-feedback signal could command a maximum of 0.5 g in addition to the nominal 5 g LIS AGCAS pitch command. Of note, if a fly-up has occurred, the pilot could assist the recovery by commanding greater than 5 g's but inputs that did not aid the recovery e.g. roll or rudder inputs were ignored.

Refer to figure 1 below for summary of a LIS AGCAS recovery.

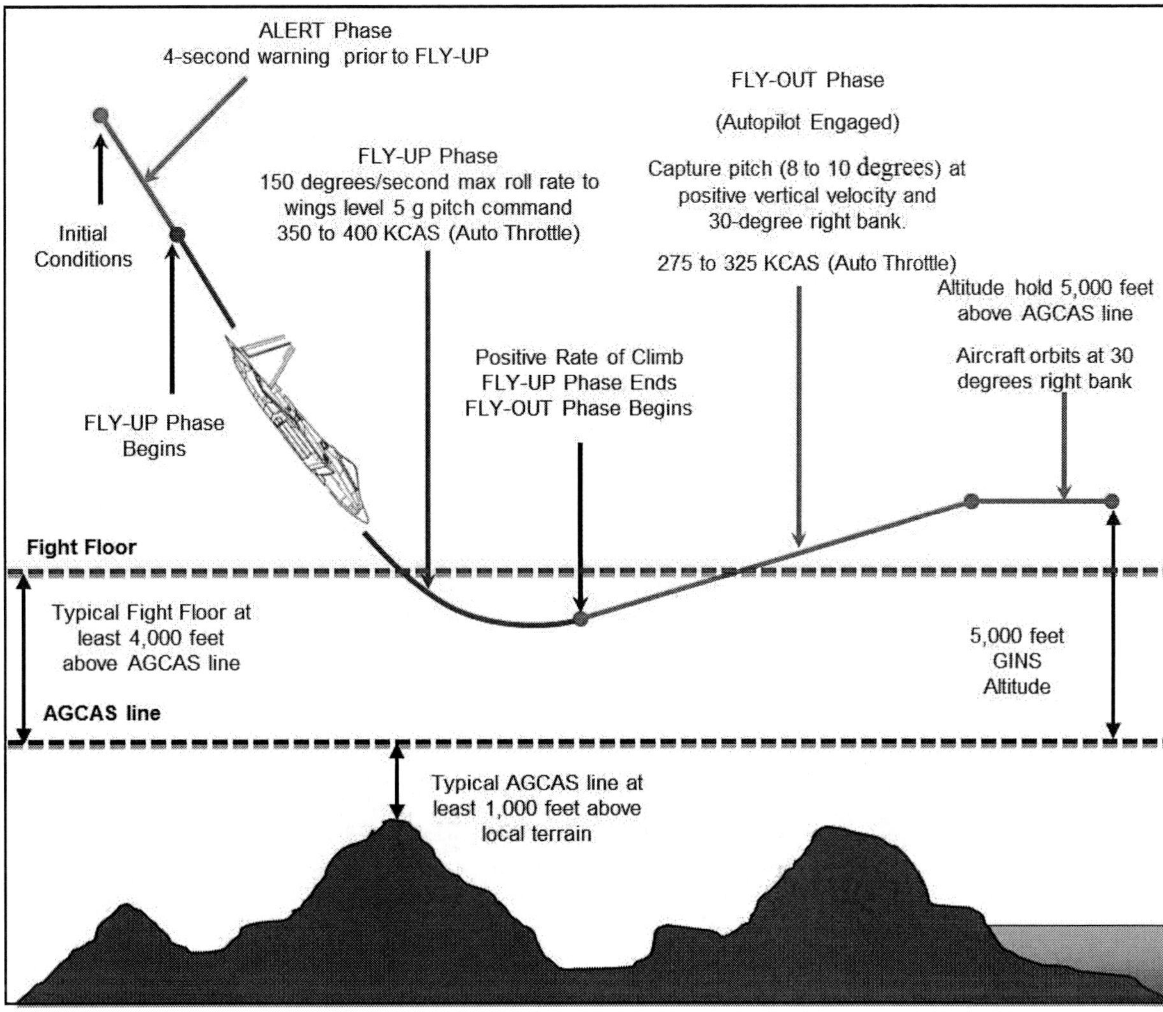

Figure 1 LIS AGCAS Recovery Summary

The LAW setting used by the legacy LAW functionality represented a barometric altitude. The AGCAS line was set by using the same procedures as the legacy LAW setting, but represented a geometric altitude with respect to the WGS 84 ellipsoid, otherwise called HAE. The LIS AGCAS used the HAE altitude determined by the Global Positioning System/Inertial Navigation System (GINS) to calculate all altitude references with respect to the AGCAS line. None of the pilot's displays had the capability of displaying the aircraft's current HAE altitude. The pilot's barometric altitude usually differs from the HAE altitude based on local temperature and pressure conditions at a given location. This difference was, on average, anywhere between 500 and 1,000 feet when testing was conducted at Edwards AFB, and could be as much as 2,500 feet at operational locations such as Elmendorf AFB, Alaska.

Interfacing with the System

The LAW readout on the HUD was located below the vertical velocity indication as shown in figure 2. The LIS AGCAS displayed the value of the AGCAS line in the HUD (via the LAW readout) when the aircraft was on the ground or when LIS AGCAS was ARMED. The readout in figure 2 shows the value of the AGCAS line (e.g., AGCAS line 10,000 feet HAE shown as "10000 LAW").

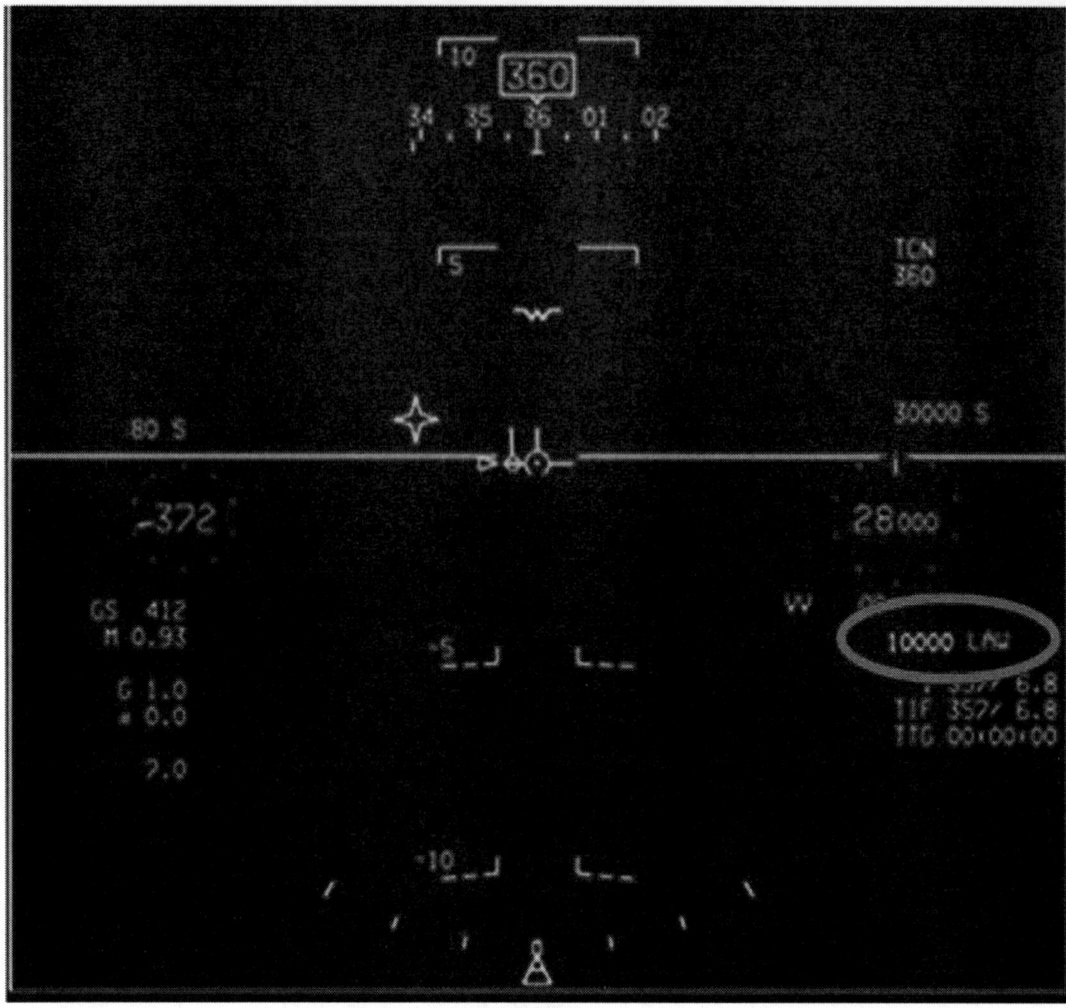

Figure 2 F-22 HUD with LAW readout

The AGCAS line was selected using the altitude function access button (ALT FAB) menu of the ICP which is located immediately below the HUD (figure 3). The AGCAS line was set by first selecting the ALT FAB (if not already selected), entering a value using the data entry keypad, and activating the menu key display labeled "LAW". Figure 3 shows the typical layout of the menu key displays when the ALT FAB was selected. Note, that the ICP and HUD still displayed "LAW" next to the numerical value of the AGCAS line in both the HUD and the ICP. A zero LAW value would turn the system off.

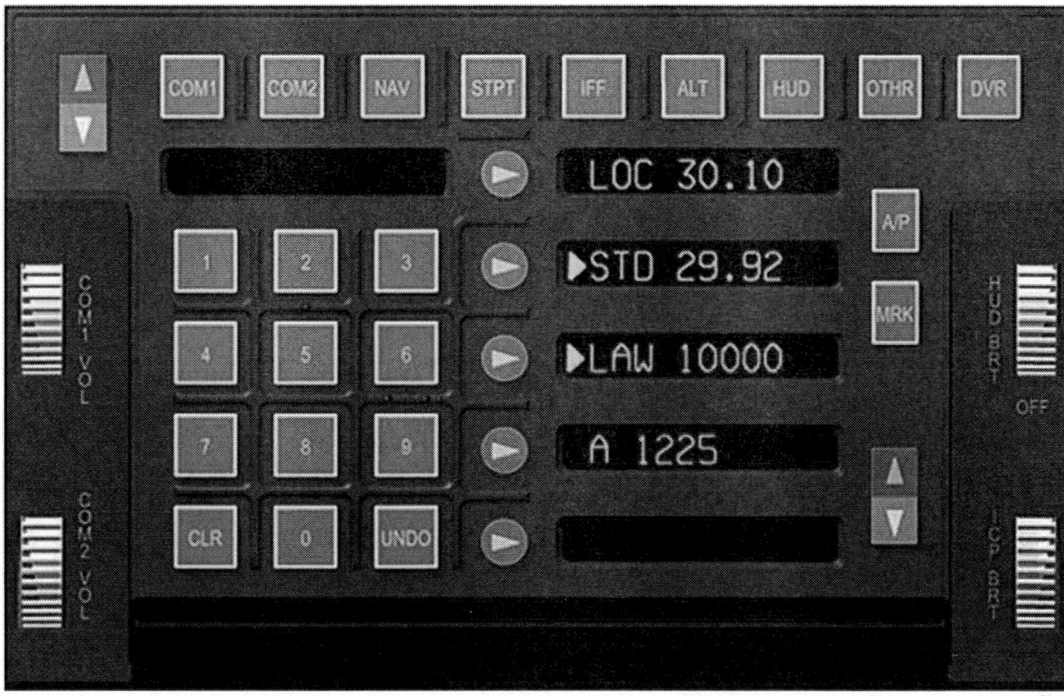

Figure 3 ICP with Typical ALT FAB Selection Display

Another way of interacting with the system was to use the paddle switch on the control stick. It is important to note that a LIS AGCAS recovery was disabled immediately when the paddle switch was activated. However, LIS AGCAS would not completely transition to a STANDBY mode until the paddle switch had been depressed for greater than one second. Different operating modes will be discussed later.

The left upfront display (LUFD), located immediately to the left of the ICP displayed all indications, caution, advisories, and warnings (ICAWs). Three ICAWs specific to LIS AGCAS are also displayed here. Every assertion of an ICAW always resulted in a warning audio to be annunciated ("*deedle, deedle*"). An "AGCAS OFF" advisory was displayed under certain operating modes (operating modes will be discussed in the following paragraphs). An "AGCAS FLYUP" caution was displayed once the *fly-up* portion of a recovery had initiated. Upon transition to the fly-out portion of the recovery, the caution would reassert and then be displayed in inverse video. A depiction of all three ICAWs is presented in figure 4.

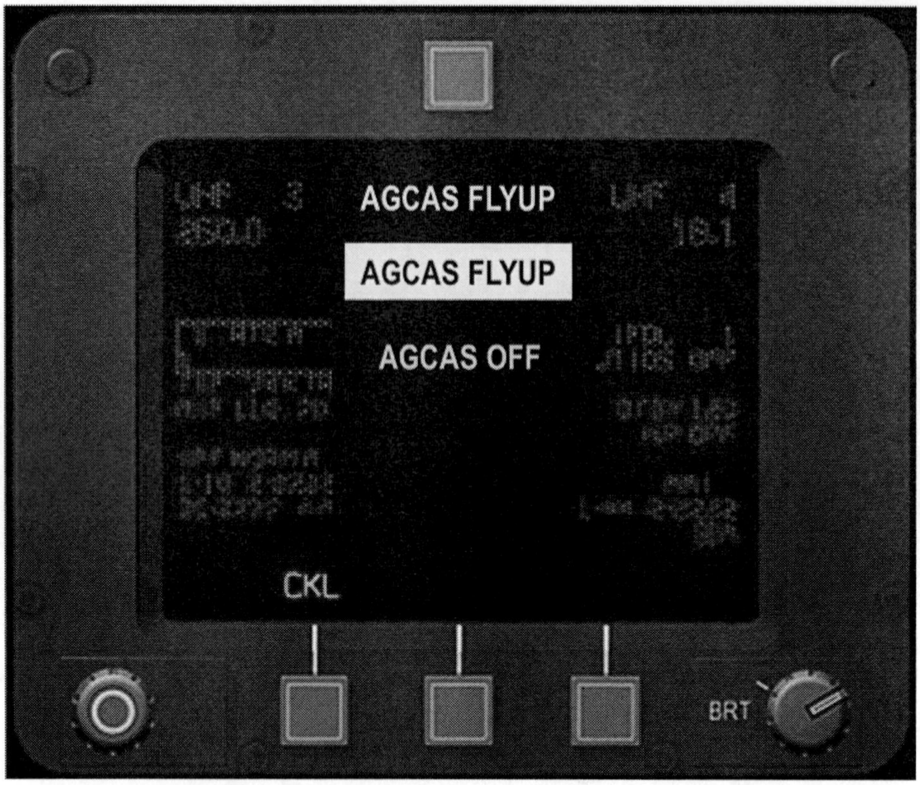

Figure 4 LUFD with Typical LIS AGCAS ICAWs
Note: Only one of these can be displayed at a time

LIS AGCAS Modes of Operation

The LIS AGCAS could only assume control of the aircraft (transition to the RECOVERY mode) to prevent penetrations of the LAW setting when LIS AGCAS was operating in the ARMED mode when TTFU reached zero. The LIS AGCAS consisted of three additional operating modes, OFF, STANDBY and DISARMED. Each mode presented with different sets of visual and/or aural cues. The OFF and STANDBY mode existed to allow the pilot to operate the aircraft without any interaction from LIS AGCAS. DISARMED mode kept LIS AGCAS from executing a fly-up if there wasn't sufficient energy or if the pilot was performing other normal tasks where a fly-up was not desired (e.g. aerial refueling or landing phase). The recovery mode (*fly-up* and *fly-out* phases) has already been discussed in detail. A basic description of the four LIS AGCAS modes is described in the following paragraphs.

OFF

The LIS AGCAS could be placed in an OFF mode when the AGCAS line was set to zero through the integrated control panel (ICP) or automatically with any LIS AGCAS related system failures. When LIS AGCAS was in the OFF mode and an AGCAS line was set, the aural tactical alert would not be issued when the aircraft penetrated the AGCAS line. When LIS AGCAS was in OFF mode, an AGCAS OFF integrated caution and warning system (ICAWS) advisory was asserted, and no LAW readout was displayed on the HUD.

STANDBY

STANDBY mode presented the pilot with the same visual indications as the OFF mode: an "AGCAS OFF" caution and absence of an AGCAS line in the HUD. However, a non-zero AGCAS line was still set in the ICP. This mode could be toggled from ARMED to STANDBY and back by depressing the paddle switch greater than 1 second. The LIS AGCAS would automatically transition to the ARMED mode if the aircraft climbed above the AGCAS line while in STANDBY. If the paddle switch was depressed again for longer than 1 second, LIS AGCAS transitioned to DISARMED if the aircraft satisfied the conditions described in the next paragraph.

DISARMED

The LIS AGCAS function was computing a TTFU but due to the operating state of the aircraft, an automated recovery was inhibited to prevent undesired fly-ups or preventing a fly-up when the aircraft was unable to perform the recovery maneuvers. The system was DISARMED when any LIS AGCAS interlocks were active (landing gear handle DOWN, alternate flaps switch to EMERG (emergency), aerial refueling switch OPEN). Additionally, the system was in the DISARMED mode when a new AGCAS line was selected and the aircraft had not transitioned above the new AGCAS line. To ensure there was adequate g available and energy to sustain the recovery, maneuvering at an angle-of-attack greater than 20 degrees or airspeeds slower than 250 KCAS also placed LIS AGCAS in DISARMED mode. If DISARMED for airspeed or AoA, "*Altitude, Altitude*" will be annunciated instead of "*Pull Up, Pull Up*" when penetrating the AGCAS line, but a recovery will not be initiated. When DISARMED, the current AGCAS line will not be displayed on the HUD, nor is the pilot notified with an AGCAS OFF ICAW. During tactical maneuvering the system transitioned from ARMED to DISARMED several times. Notifying the pilot of every switch would be too distracting. The pilot must be cognizant of his flight regime and understand those times when CFIT protection is not provided. It was demonstrated when operating under normal Air Force training rules with the AGCAS line set 1000 feet above the terrain, the system would typically ARM before the AGCAS line and recover the aircraft.

ARMED

The aircraft was ARMED when above the current AGCAS line, within the specified maneuvering envelope, and there were no interlocks present. The pilot was alerted of the ARMED status when the current AGCAS line was displayed in the HUD. CFIT protection was being provided and the system was ready to take control of the aircraft to prevent penetration of the AGCAS line.

TEST/SAFETY PLANNING, PREPERATION, AND EXECUTION

This paper presents information from two different developmental test phases. The first developmental test phase involved running a select number of test points that intentionally provoked a recovery to initiate. These test points (approximately 200) were used to validate the recovery performance model of LIS AGCAS developed by Lockheed Martin. This test phase lasted from July to December of 2013. Very little military utility testing was afforded during this test phase. The system was considered to be unproven at this point. Operations were restricted to those with control room monitoring only in pursuit of dedicated performance recovery test points. Very little information could be inferred with respect to LIS AGCAS interference with normal operations. The second test phase was concurrent with the fielded version of the

LIS AGCAS algorithm included into the Update 5 OFP and allowed testing to be conducted in more operationally representative scenarios. The follow-on testing of LIS AGCAS in the Update 5 OFP occurred from June 2014 to November 2014.

Due to the inherent man-in-the-loop architecture of the system, aircrew had concerns before testing even began. Some of the concerns were:

- Changing the line due to changing terrain in changing airspaces
- Simply forgetting
- Entering incorrect values
- Misunderstanding the state of the system
- Close proximity to other aircraft (battle damage checks, formation flying, tactical engagements, instrument flying)
- Desensitization

Near the end of this test phase, pilots were allowed to exercise the system with a little more freedom to explore and assess these concerns. Less scripted BFM and basic intercepts were executed to assess nuisance potential in the larger mission set of the F-22. As expected, it was found that the higher the AGCAS line, the more it would interfere with normal F-22 tactics. The mission focus of the F-22 does not afford the pilot the luxury of focusing on the state of AGCAS at any given moment. The CFIT protection offered by LIS AGCAS would be most crucial when engaging lower altitude threats; therefore, it is recommended that the AGCAS line be set as low as practical for the area of concern to lower chances of unnecessary fly-ups but still provide some CFIT protection. The overall assessment could not be determined because the less scripted BFM and basic intercepts were developmental military utility evaluations and not end-to-end system operational tests.

Maneuver Development

The AGCAS test points required specific set up conditions to achieve the desired test conditions (altitude, load factor, Mach, bank angle, dive angle, etc.). This was very similar to a Loads maneuver; therefore, the existing F-22 Loads Maneuver Develop Process was used to achieve the initial set up conditions for each AGCAS test point. This required extensive simulator practice during the development of the test cards for each maneuver.

The test team had multiple simulators available. The only simulators that were capable of running the LIS AGCAS algorithm during the developmental test phase were located at the Lockheed Martin facility in Fort Worth, Texas. To develop aircrew familiarity and initial setup conditions, a majority of test points were practiced in the Fort Worth simulators. Simulators at Edwards AFB were used to refine setup procedures by attempting to attain the required recovery entry conditions at an altitude at which LIS AGCAS was predicted to engage. Actual LIS AGCAS recovery algorithm performance could not be assessed in the Edward's simulators.

The developmental test phase of LIS AGCAS required considerable safety planning. The chief safety planning metric used for the recovery maneuvers was Time Safety Margin (TSM). TSM was developed following the fatal F-22 CFIT accident in 2009 and is defined as the time in seconds to directly travel from the worst-case vector to an unrecoverable condition. Calculation methods for TSM vary from very generic, aircraft agnostic, dynamic equations to complete aircraft-specific 6 DOF simulators. The test team used the F-22 ATLAS model to calculate the TSM for every planned intentional LIS AGCAS recovery; however, the local F-22 CTF flight sciences simulator did not have the LIS AGCAS capability.

During the third test flight of LIS AGCAS, a particular test point was attempted with the following expected engagement conditions: 1.5 Mach / -85 degrees dive angle / AGCAS line: 12,500 feet / expected engagement altitude at 31,440 feet. This test point had a calculated TSM of 10.9 seconds. The test pilot did not believe LIS AGCAS had engaged in a timely manner and aborted the test point due to differences between system HAE and displayed pilot's barometric altitude. This difference in barometric and system altitudes caused the pilot perceive that he was approximately 1,000 feet below the altitude that LIS AGCAS was expected to engage. LIS AGCAS had in fact engaged and was performing a recovery, the pilot's aft stick inputs improved the performance of this recovery and the aircraft recovered at approximately 18,000 feet, 6,500 feet above the AGCAS line setting of 12,500 feet.

The test pilot was further frustrated by the amount of ground rush during the abort and felt that if he waited the full 10.9 seconds before aborting, as defined by the TSM calculation, the aircraft would have been unrecoverable. The TSM calculation was correct as defined in the Air Force Instructions. TSM was designed as a tool to alert the test team of higher risk test points and was never intended to be the sole metric for risk determination. It was not fully appreciated how this highly dynamic point and the performance of the F-22 rapidly worsened the dive parameters post test and significantly reduced the perceived recovery performance. This led the test team to re-evaluate the value of TSM.

While it was determined there were no issues with the TSM calculation, the test team re-evaluated the value of TSM as the sole metric for risk assessment. The TSM calculation is valid only for the point in time at which it was calculated, and assumes the pilot begins a dive recovery immediately when any of the abort criteria are met. The previously mentioned test point was a highly dynamic maneuver; the aircraft was in a near vertical dive, at maximum thrust, and accelerating. While the methods used to calculate TSM for AGCAS accounted for dives with elevated g, it did not account for changes in abort velocity or flight path angle. For this test point, an unrecoverable state would be reached with a delay in recovery by 4-5 seconds despite a 10 second TSM.

It became clear that team members had varying understandings of TSM. As a result, the test team implemented an internal policy to perform a Post Abort Non-Responsive Test (PANT) check in the F-22 6-DOF flying qualities simulator for every test point. This check required the pilots to fly the entire maneuver to recovery from the abort conditions defined by the TSM calculation. If at any point there was doubt (normally based on ground rush or pilot judgment), then the point was re-accomplished with an added delay of 4 seconds past the abort conditions allowing for accelerations and/or worsening conditions of the TSM calculation before performing a manual recovery to ensure ground avoidance. During this delay, the pilot continued to execute the test point maneuver with the goal of aggravating the state of the aircraft (make adverse inputs to degrade the aircraft flight path). If ground clearance was not possible with this 4 second delay, the test point conditions were adjusted (e.g. moving to a lower ground elevation, raising the AGCAS line and subsequent abort altitude, or more restrictive abort conditions on dive, bank, and Mach). By revisiting each maneuver in the simulator, the test team was able to better understand the dynamics of the maneuver and characterize any unexpected behavior during the dive recovery for the remainder of the developmental test phase.

Pilot Vehicle Interface

The developmental test phase proceeded without further incident with respect to the recovery performance. However, it became apparent that the inherent design of the system (LIS vs. DTED) made the system susceptible to many nuisance warnings and the PVI was deficient. On the second flight of the developmental test phase, a test pilot forgot to change the AGCAS line after completing the final test point planned for that sortie and preceded to Return to Base RTB. LIS AGCAS still being active and in the armed mode, annunciated the "*Altitude, Altitude*" ("*Altitude, Altitude*" was used in the initial release of LIS AGCAS instead of "*Pull Up, Pull Up*") and dangerous condition tone to warn the pilot of an impending recovery as the aircraft descended towards the AGCAS line. The pilot was accustomed to hearing the "*Altitude, Altitude*" and simply ignored the warning during RTB because of its use with the legacy LAW functionality. A recovery initiated before the pilot could disarm the system. In this case, the chase aircraft had just been directed to rejoin formation with the F-22 but was fortunately still approximately a mile away when the recovery had occurred. This could have led to a mid-air collision if the F-22 had been in closer proximity to the chase aircraft.

This unintentional recovery prompted two very important changes. The amount of warning time when the aircraft is in the benign maneuvering envelope was extended from 4 seconds to 6 seconds to afford pilots more time to assess their situation and act appropriately (e.g. maneuver away or "paddle off"). Most importantly, the "*Altitude, Altitude*" was changed to "*Pull Up, Pull Up*", a more directive tone that was not previously associated with any other function of the aircraft,. The extended warning time was implemented before the end of the developmental test phase, whereas the more directive warning tone wasn't incorporated into LIS AGCAS until the Update 5 OFP (fielded version).

These changes helped pilots recognize impending LIS AGCAS recoveries more easily; however, they did not improve overall situational awareness of LIS AGCAS operating state at all times. The unintentional recovery mentioned above occurred even in the presence of procedures clearly documented on the test card to "Set LAW=0" effectively turning the system to the OFF mode. Furthermore, the LIS AGCAS and AGCAS line was monitored from a mission control room. Human error is alive and well.

The developmental test phase presented multiple scenarios where a task saturated pilot required valuable time to perform a mental cross check of the LIS AGCAS operating state. Determining the operating state required referencing up to three different locations in the cockpit (HUD, LUFD, and ICP). A table of these locations as well as the indications (or absence of indications) is included Table 1 below.

Table 1 Basic Cross-Check of Visual Cues for Discerning Basic Modes of LIS AGCAS

LIS AGCAS Mode	AGCAS line Present in HUD?	"AGCAS OFF" caution Displayed?	Zero or Non-zero AGCAS line in ICP
OFF	NO	YES	Zero
STANDBY	NO	YES	Non-Zero
DISARMED	NO*	NO	Non-Zero
ARMED	YES	NO	Non-Zero

Notes: 1. * If in the DISARMED mode AND on the ground, the AGCAS line was present in the HUD.
2. Transition to OFF or STANDBY was marked with a "Deedle, Deedle".
3. Transition from OFF or STANDBY was marked with a parking brake tone (in flight).

Determining the operating state of LIS AGCAS was challenging at times. The most straight forward case was when the system was in the ARMED mode because the pilot would only have to reference the HUD and observe the LAW readout. If the LAW readout (see figure 1) was visible, then not only did the pilot know that the system was armed and providing protection, but was also aware of the current AGCAS line setting. There could be several reasons why the LAW readout might not be present in the HUD:

1.) DISARMED mode (AoA greater than 20 degrees, or slower than 250 knots)
2.) STANDBY mode because the pilot activated the paddle switch for greater than 1 second earlier in the flight
3.) Off
    a. Pilot commanded (LAW 0 set but not displayed in the HUD)
    b. System commanded because of a LIS AGCAS related failure

Determining the operating state of LIS AGCAS could require looking in three separate locations for a combination of the presence and absence of cues. This was challenging and probably not the best method to convey the state of an "automatic" system. This implementation is not consistent with human factors engineering guidelines DOT/FAA/CT-03/05 HF STD-001, *Human Factors Design Standard*, and MIL-STD-1472G, *Department of Defense Design Criteria Standard: Human Engineering* (references 2, and 3), which state that automation "should not increase demands for cognitive resources" and that the user should be "appropriately informed to maintain awareness of the situation and other status of automated functions".

Nuisance Potential

There are numerous reasons for nuisance recoveries when operating with a LIS based AGCAS. First, when operating in mountainous terrain the highest terrain may be above required operating altitudes. Typically the AGCAS line is set approximately 1,000 feet above the highest terrain; however, the aircraft may still be required to operate at lower altitudes in the valleys. These operations essentially require the pilot to turn LIS AGCAS off and operate without CFIT protection when it is needed the most. Second, misunderstanding the appropriate AGCAS line setting for the local area of operations. Third, incorrect altimeter or LAW settings, and finally just forgetting to reset the AGCAS line appropriately.

A more disturbing consequence of the nuisance warnings is the concept of a pilot becoming desensitized to these warnings and preventing a LIS AGCAS recovery that is actually required. Nuisance warnings and unintended fly-ups will cause pilots to trust their instincts rather than LIS AGCAS. This defeats the whole purpose of incorporating LIS AGCAS in the F-22. As a result, it is recommended that the line be set as low as practical and Concept of Operations (CONOPS) be developed that require minimal interaction to prevent the nuisances and interactions listed above. This behavior was demonstrated during the developmental test phase when a test pilot accidently set an altimeter of 32.00 inches of mercury. The pilot was attempting to set an AGCAS line of 3,200 feet but accidently changed the altimeter to 32.00 inches of mercury. Refer to figure 3: the number entry on the ICP for both 32.00 inches of mercury and 3,200 feet are identical. Altimeter settings and AGCAS lines are both manipulated with the ALT FAB and separated by only one line on the ICP display. The pilot was unaware that his aircraft was actually flying approximately 1,000 feet lower than indicated. LIS AGCAS alerted the pilot of a need to initiate a recovery which the pilot quickly recognized, assumed the system was incorrect and "paddled off" LIS AGCAS, preventing the fly-up. This scenario represents a situation where the pilot has become desensitized to assumed nuisance warnings.

Recovery Performance

Aside from PVI issues, the ability of LIS AGCAS to recognize when a penetration of the AGCAS line was imminent and recovering the aircraft above that line was excellent. During the developmental test phase, 178 recoveries were performed to assess the performance of the LIS AGCAS algorithm, 51 of these recoveries did penetrate the AGCAS line. These penetrations were the result of aggressive maneuvers near the corners of the operating envelope and were intended to validate and update Lockheed Martin's performance model. A brief statistical breakdown of these recoveries is included in the Table 2 below (reference 4)

Table 2 Recovery altitude deviation performance summary

Recovery Altitude Deviation Extremes			
	Clean	External Tanks (2)	Combined
Minimum recovery altitude deviation	-410	-360	-410
Maximum recovery altitude deviation	810	1,370	1,370
Average recovery altitude deviation	130	300	190
Standard Deviation	250	370	320
Recovery Altitude Deviation Summary			
RAD ≥ 0	64.0%	80.8%	70.8%
$-200 \leq$ RAD < 0	31.6%	15.4%	25.0%
RAD < -200	4.4%	3.8%	4.2%

Notes: All recovery altitude deviations (RAD) are in feet. Negative numbers represent recoveries where the aircraft recovered BELOW the AGCAS line

SUMMARY AND LESSONS LEARNED

The recovery performance of LIS AGCAS was satisfactory and met design requirements. The LIS AGCAS was effective in determining when a penetration of the AGCAS line was imminent and recovered the aircraft. The LIS AGCAS is only effective in preventing CFIT mishaps if an appropriate AGCAS line is selected by the pilot and the aircraft is within the LIS AGCAS operating envelope (greater than 250 KCAS and less than 20 degrees angle-of-attack [AoA]).

The pilot-vehicle interface (PVI) was marginal. Attributes of the visual cues and aural tones decreased situational awareness (SA) and increased workload. The pilot perceives the ground as the threat whereas LIS AGCAS avoids a pilot-entered arbitrary line. Visual cues and aural tones were ambiguous in determining LIS AGCAS operating state and increased pilot workload. At times the pilot was required to reference three different locations to determine the operating state. In addition, the absence of cues led to misinterpretations of the LIS AGCAS operating state. This could increase the potential for unexpected fly-ups or decrease CFIT protection. The increased pilot workload associated with LIS AGCAS distracted from other important tasks. The extra attention required to properly set the AGCAS line, maintain situational awareness on the current operating state of LIS AGCAS, and maintain awareness of the aircraft's position relative to the AGCAS line may not always be practical during complex operations. Based on flight test experience, it is likely that operational pilots will periodically make errors when setting the AGCAS line, have degraded awareness of the LIS AGCAS operating state, and not be cognizant of aircraft position in relation to the AGCAS line.

It was demonstrated during testing that even with trained, familiar and AGCAS-experienced test pilots that the PVI requiring constant pilot attention resulted in AGCAS operation errors including one accidental fly-up and numerous fly-up warnings. During Update 5 testing, it was observed that at least 28% of the time, pilots would forget to change the AGCAS line as required by the changing terrain (reference 5). It was also observed that approximately 50% of the time during a BFM engagement, the pilot was operating outside the AGCAS envelope and had no CFIT protection. In many tactical scenarios when engaging threats in the low altitude environment, the pilot is statistically more likely to hit the ground. During these scenarios, the pilot is required to simultaneously conduct dynamic maneuvering, while interpreting data "heads-down" with the aircraft pointed at the ground. A distracting LIS AGCAS warning or the added PVI required to change the AGCAS line could be the last straw that may result in a CFIT accident. Most likely, pilots will simply turn LIS AGCAS off in these situations.

The excessive pilot interaction required by the F-22 LIS AGCAS PVI design could result in system operation errors that could cause unacceptable unintentional fly-ups. These unintended fly-ups pose a safety concern primarily of two distinct types:

- Close proximity with other aircraft such as formation or battle-damage checks and
- Unintended fly-ups in controlled airspace during IMC flight. If any one of a number of human errors occurs (e.g. forgetting to reset the LAW, incorrect setting, incorrect altimeter setting, etc.), a fly-up into controlled airspace during IMC could be potentially disastrous. At a minimum this would result in the F-22 pilot declaring an immediate IFE with the controlling agency during the fly-up and pose an unnecessary risk to aviation safety

PVI design issues and system implementation are often the result of misguided direction or poorly defined requirements. The old adage of "If you want it bad – you'll get it bad" may apply somewhat to the F-22 LIS AGCAS design. The USAF certainly had the right idea when directing the implementation of an AGCAS system to reduce CFIT accidents especially after two fatal F-22 CFIT accidents; however, implementation of this direction may have been short sighted. Lockheed-Martin certainly designed an excellent algorithm that was demonstrated during flight test to perform very well; however, the PVI aspects and required pilot interaction ultimately make an excellent system appear to be marginal.

Early in the requirements phase, F-22 test pilots highlighted that a DTED-based AGCAS system was preferred and need not be a dual-redundant flight critical system. The F-22 SPO directed that any AGCAS system on the F-22 would be a flight critical system. Based on the computer architecture of the F-22 this would have been a serious re-design of the computer architecture with exorbitant costs. The F-16 AGCAS system is DTED-based, is now fielded and credited with at least two confirmed "saves." It is a single-strand system rather than dual-redundant and was developed just prior to the F-22 AGCAS development but still quite relevant to the F-22 effort. The F-16 uses the System Wide Integrity Monitor (SWIM) to continually check the FLCS/AGCAS and either allows the AGCAS to command a fly-up or turns the system off for various system failures. SWIM is the same system that was utilized successfully for years during Low Altitude Navigation, Terrain Following Infrared for Night (LANTIRN) operations at night – certainly flight critical! AGCAS is already a backup system and the requirement to be a dual-redundant flight critical system is a poorly stated requirement that ultimately had a negative impact on AGCAS implementation on the F-22. It is not clearly understood by our team why the requirements levied on the F-22 program were so divergent from the F-16 and did not leverage lessons learned from their highly successful program. Perhaps communication, education, and utilization of the proper chain of command versus program channel communication are all potential reasons for this outcome. In any event, one clear lesson learned is that proper communication of all details of a potential design to include background from other platforms is essential to avoid poor definition of requirements and potentially fielding a less than optimum design for the budget available.

It is equally important to note that test pilots alone have the unique position of having experience in multiple platforms in the environment of intended use, understanding the engineering aspects (since we are all engineers), and the ability to communicate with our ultimate customers on an equal footing. Engineers, program managers, budget analysts and cost accounting managers do not have the benefit of our operational experience, applied engineering knowledge as well as having the breadth and depth of knowledge and powers of observation that test pilots tend to acquire after their formal training. As test pilots we need to take the lead in ensuring proper communication with our customer as well as inserting ourselves early in the requirements process to avoid potentially fielding a less than optimum design. Our job is to ensure that our customer can always "Fly and Fight!"

Ultimately with extensive pilot training to overcome PVI issues as well as thorough and appropriate CONOPS, the F-22 LIS AGCAS will save lives and prevent CFIT accidents in the F-22 fleet.

References

1. AGCAS Tiger Team, *Automatic Collision Avoidance Technology/Fighter Risk Reduction Project: Auto GCAS Return on Investment Assessment*, September 2007. AGCAS Tiger Team report to ACC/A8

2. DOT/FAA/CT-03/05 HF-STD-001, Human Factors Design Standard (HFDS) For Acquisition of Commercial Off-The-Shelf (COTS) Subsystems, Non-Developmental Items (NDI), and Developmental Systems, U.S. Department of Transportation Federal Aviation Administration Headquarters Human Factors Division, Washington, DC, May 2003.

3. MIL-STD-1472G, Department Of Defense Design Criteria Standard: Human Engineering, USAF Department of Defense, Washington, DC, 11 January 2012.

4. 412TW-TR-13-42, F-22 Line In The Sky (LIS) Automated Ground Collision Avoidance System (AGCAS) Test and Evaluation, 412 Test Wing Edwards AFB, California, January 2014.

5. 412TW-TR-14-40, F-22 Update 5 Operational Flight Program (OFP) Air Vehicle Systems Test and Evaluation, 412 Test Wing Edwards AFB, California, March 2015.

Integrated Night Vision and Testing the Striker II HMD

Peter Kosogorin

BAE SYSTEMS

The role of the attack pilot is afforded a significant degree of conventional stealth through operation in periods of darkness. The UK and coalition allies are more likely to be able to operate on Day One of a conflict or operation if, amongst other things, they have a credible night capability. Electro optical and enhanced image intensification technology has provided night vision to aircrew and attack pilots for decades, but only recently has fully integrated night vision become a reality. On the back of the F35 development programme digital helmets have been designed and now offer integrated night vision (INV) to today's aircrew. No-one can deny the importance of this night capability, but what is also undeniable are the increased hazards associated with flying a high performance aircraft at night and particularly at low level. The risk of CFIT increases at night when terrain or obstacles are not clearly visible and the risk of collision with other aircraft also increases. To furnish the capability whilst mitigating these risks is a difficult task, one in which integrated night vision would appear to excel.

There are numerous clear advantages of integrated night vision, the first of which is the physically obvious: by removing the requirement for heavy NVGs to be attached to the front of the helmet the comfort of the aircrew is increased. Mass is reduced and the CG solution is more balanced allowing for longer sortie duration with commensurate reduction in fatigue levels. The pilot has a wider field of view, can more easily turn his head even under increased g forces with no canopy interaction issues and there is no peripheral obscuration caused by the NVG tubes and mechanism. Using HOTAS control there can be a seamless transition from day into night and night into day and if the image-intensified video is not optimal it can be instantly deselected. There is no requirement to store, fit and adjust heavy equipment including the face protection visor. This means reduced workload for the pilot – no two-handed donning and doffing procedure. The system does not require jettison and albeit having a slighter higher mass than simple helmets, it does not pose an increased hazard on ejection. One helmet is all that is needed and the aircraft symbology is also always available as is not the case with many legacy systems. The night vision camera (NVC) can also be day capable, providing a recording system to aid in training and mission debrief. Perhaps the most exciting capability offered is the potential ability to mix any sensor video in the display such as targeting imagery or dynamic synthetic vision.

The Striker II Helmet was conceived out of the requirement for a replacement F35 HMD owing to the incumbent's risk. The first iteration was required to be an NVG solution owing to the risk of NVC technology at the time but after one year it moved to the INV solution. The helmet reached maturity but in 2013 was no longer required by the programme so became self-funded.

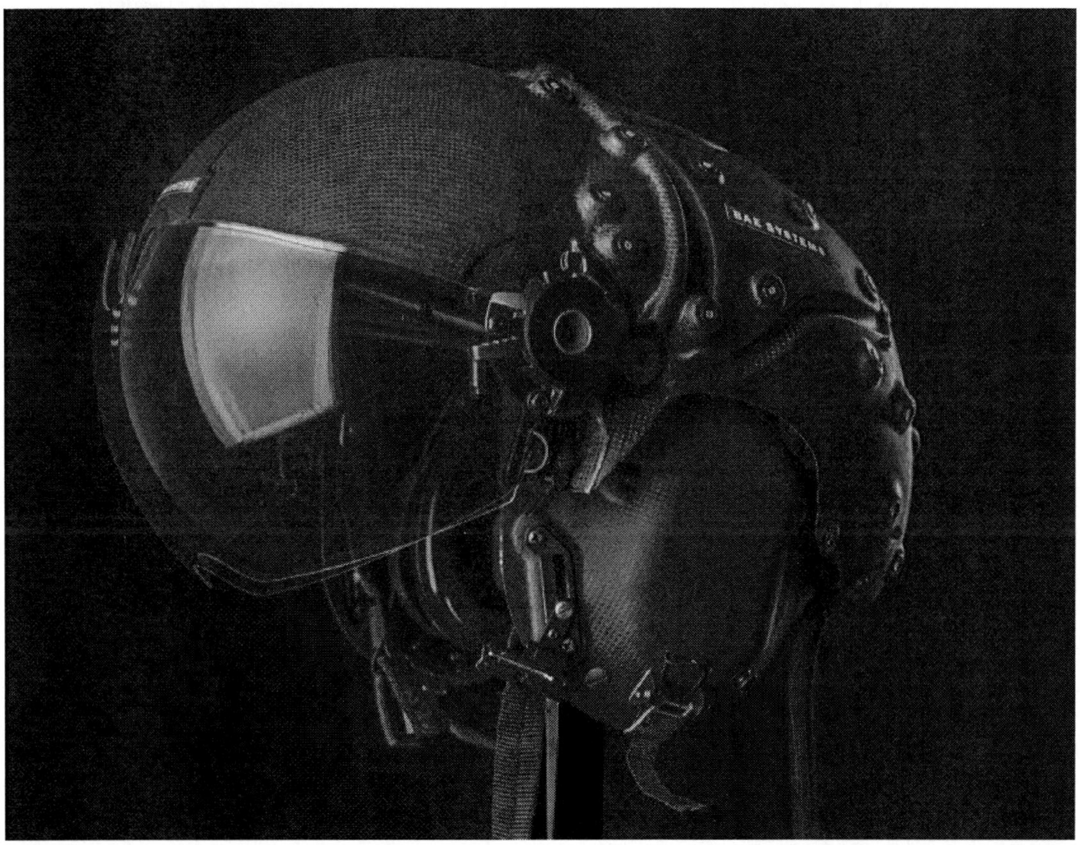

The Striker II helmet has a strong heritage in the Striker I HMSS (Helmet Mounted Sighting System) or HEA (Helmet Equipment Assembly) and retains some of the earlier features of the NVG-equipped solution such as nape and ear cup tensioner adjustments. It also retains the legacy flight clearance limitations in KCAS, altitude and g. The blast/display visor has a greater transmissivity than legacy at 76% and the total equipment weighs only 4.84lbs. It consists of 3 main parts starting with the individual custom fitted protective liner. The inner helmet carries the communications system and comes in 3 sizes – this connects to the Advanced Dynamic Oxygen Mask. The one size outer helmet assembly includes the NVC, image processor, tracker LEDs and visors. The Helmet Equipment Assembly Service Module (HEASM)

is the man-portion of the electronics unit and attaches to the ejection seat. Production variants will have the HEASM left in the aircraft and a simpler and lighter cable connection increasing comfort and simplicity for the user.

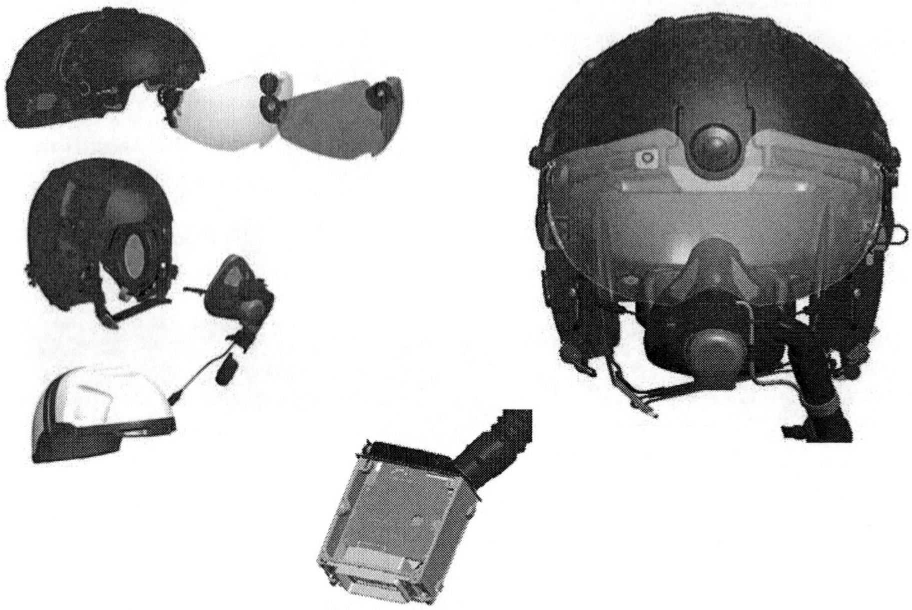

At the heart of the INV on the Striker II is the INTEVAC ISIE 11 Electron Bombarded Active Pixel Sensor (EBAPS), which provides image intensified video with low noise and 1640 x 1240 pixel resolution. It has a significantly reduced scintillation noise compared to NVGs and it is equipped with Automatic Gain Control. The Low Halo camera increases usability in conditions of increased cultural lighting and an integrated Flash memory allows storage of image correction parameters.

The Striker I symbology includes an Arc Segmented Attitude Reference and nominal flight and tactical data. Striker II symbology is under development and will be similar, but for the test involved a fixed set only. Symbol font changes were investigated during the trial including black casing, which on the video projection was expected to provide the optimum solution. The 'excess' NVC resolution was used to provide a Zoom capability, which provided a picture in picture display of a 1.6x magnified image on boresight while the button was selected. Ostensibly this capability was designed for the use of identification of ac and targets to assist in fulfilling ROE requirements.

The helmet's opto-inertial tracking system consists of 10 clusters of 4 Infra-Red LEDs mounted on the Outer Helmet Assembly. The inertial sensor is inside the Helmet and is designed to improve dynamic performance, head box coverage and compensate for aircraft buffet on some platforms. There were 3 cockpit mounted sensors for the trial but for future integration of the Striker II the plan is for only one sensor, located on the head box. Head position and orientation is determined by the best LED cluster and sensor combination. Each sensor optic has a wide field of view (nominally 80° x 60°). There is 100% coverage even if optical tracking is lost and with no magnetic element, there is no cockpit mapping – the tracker is fully tolerant to cockpit changes. A body pack controller of the system was utilised for the trial, which was fitted to the right side of the LSJ stole. This equipment was incorporated for trial purposes only but allowed freedom and flexibility in control of the NVC and symbology settings. It allowed us to select NVC and Symbol Optics as well as other parameters in order to make a more comprehensive assessment. It was essential for the trial but would be overly complex for production. Full HOTAS control will be integrated for the production version.

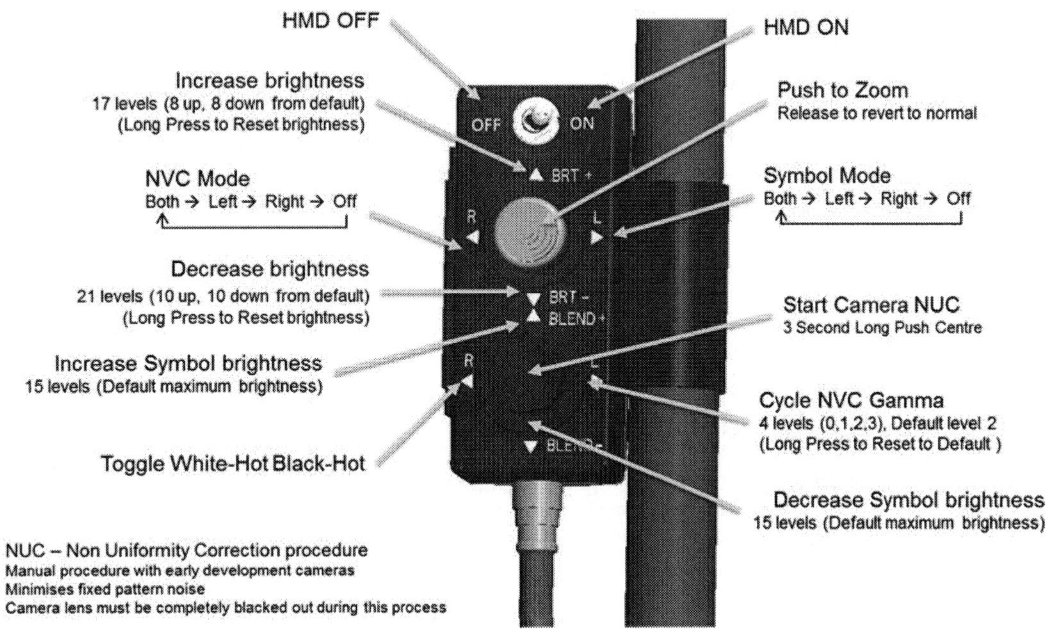

The platform chosen for testing was a two seat Tranche 1 Instrumented Production Typhoon trainer, IPA1. We were forced into using a legacy clearance for the helmet, based on the previous NVG-equipped solution which provided us with a more restricted envelope than would have been ideal. We had an unambitious initial plan to simply test the technology and comment subjectively on the NVC performance, operational acceptability of the helmet and its impact on the cockpit environment on the ground and in the air. We needed to investigate the intended improvements in terms of comfort and fit and the performance in a range of ambient conditions, including overcast and moonlit conditions. We planned a minimum of two subject pilots for cross section, and flew four with five sorties total. We were very fortunate that nothing was lost to weather or serviceability as we had very little contingency time.

The performance assessment included gauging and assessing the effect of the introduction of the NVC on the aircrews' visibility of the external environment, cockpit switches, controls and displays in a dynamic lighting environment including camera image Parallax error and field of view. We looked at the optical performance of the DHA visor with and without the NVC blended image looking both through the HUD symbology / external scene using all channels, (Both, Left & Right). We also assessed the clarity of the symbology both with and without NVC image (again B, L and R). In terms of NVC performance we looked at Image acuity or clarity including halo effect, resolution, sharpness, contrast and the effects of various levels of cultural lighting. We examined the effects of cockpit displays and potential in-cockpit reflections, against the prevailing conditions encountered. We wanted to quantify perceived changes in pilot workload and mental fatigue whilst wearing the DHA with the NVC image present. We had fully anticipated some issues caused by parallax owing to the 'cyclops' position of the sensor but would having a monocular sensor affect depth perception or would eye strain be experienced by the pilots?

Bearing in mind the relatively unambitious objectives of the trial the final sortie profile seemed rather complex:

- NVC OFF Take Off
- Helmet 'familiarisation' – subjective assessment
- Medium Level Navigation exercise in mountainous and flat terrain.
    - Direct Comparison with NVG – various cultural lighting levels
- Formation
    - Conventional – Radar Intercept to VID
    - NVC ON
- Close Air Support
    - Air – Surface Weapon Profiles
    - Work with dedicated Ground FAC+ IR illumination
    - Integral use of Litening III Pod
- Radar Approach to overshoot Dark Airfield
- Land NVC OFF

We had identified a number of hazards, including NVC failure leading to display bright up, display image freeze, disorientation and comfort issues. These were mitigated mostly through the second pilot although we did not want to limit ourselves to this as a requirement. Other mitigations included selecting the NVC off <200ft, an incremental approach to formation and a ground assessment prior to the night trial.

There were a number of factors that affected the performance of the system during testing. In the UK the timing of night trials is very important. We planned for November, with a last landing time of 2300 (owing to local noise regulations) providing us with a great deal of sortie flexibility and time, with darkness falling at around 1800. Unfortunately, by the time we got through all the paperwork and clearances we were to fly in May. With all of 20 minutes of true night to take advantage of for each flight. Whilst we wanted to assess the equipment in a variety of conditions we really did need more time in full darkness and most of the sortie was executed with a good red glow on the horizon. We had considered parallax and latency but not the effect of the NVC image overlaying a strong visual image and other issues were significantly affected by this. This effect, exacerbated by the ambient lighting conditions, swayed the results. The Non Uniform Correction (NUC) procedure is required at each NVC start up and is an involved process and guidance was not forthcoming. The NUC procedure had to be carried out by placing the pilot's gloved hand over the lens for the duration and there is no real feedback of success. We were informed that - *If NUC is procedure not carried out the NVC image will be significantly degraded at this standard and considered to present an unusable standard of image.* It was deemed unreliable such that if any light escaped past your gloved hand results could be poor.

The results of the trial were generally very good. The digital system worked, and whilst most operations currently undertaken with NVGs could be undertaken with INV there were very few areas where NVGs would probably remain the chosen solution. Test symbology was extremely clear and the black cased symbology was optimal. Comfort levels, head motion and FOV, in particular the ability to look down and behind the 3/9 line, were excellent although there was one comment concerning obstructed movement due to cabling. We reached 5.5g and pressure breathing caused no issues with full head movement and image visibility. Workload was significantly reduced in the utilisation of Night Vision equipment and the simplicity in transition from Day into Night was a breath of fresh air. Parallax was not an issue at all but the image overlay effect garnered some negative comment and caused some exacerbation of latency in the NVC projection. This effect was described by the RAF Test Pilot as when *"...Looking towards the sunset reminded me of a 1980s David Bowie LSD Trippin' Music video like image......"* Other pilots corroborated this view particularly when looking at cumulus cloud towards the set sun. In close formation for example, the target aircraft and ELPs could be seen with the naked eye, whilst the NVC gave an image that was slightly mis-matched. When there was head movement on top of this the image caused a little momentary confusion for the pilots. There were no eye fatigue issues with viewing the display and depth perception was not an issue either on the ground or in the air. Distance could be judged when close in but perception was similar to that when using NVGs when further out. The zoom function was a good idea but the acuity of the system was such that when viewing the zoomed image it was not always possible to break out more detail.

A list of recommendations collated by all pilots involved in the trial was made and published. The NUC needs to be reliable, effective and intuitive. If this is such a critical procedure there needs to be no doubt that an effective correction has been executed. There are plans for a shutter to be fitted to the lens which will help in its reliable execution. The image definition was not to the same standard as NVGs and NVGs performed better with the FAC IR Target Marking equipment. This is where the subjective assessment becomes critical. The image is acceptable in most environments and the ease of use outweighs the image quality in all but a few areas of flight, making INV the equipment of choice over NVGs. Whilst the ISIE 11 provides a high resolution image the processor and projection systems cannot display the full quality on the pilot's visor. Even though we have a 'super-sophisticated system', the cabling attracted negative comment and for two pilots obstructed head movement and was almost found unacceptable. This was owing to a corrugated cabling set up close to the helmet and should have been discovered on ground testing or in flight during the day first. The visor lock caused one pilot some issue and the glare visor requires a more robust attachment. Comfort is critical to many customers and these issues, whilst appearing minor, should not be underestimated. The flight limitations imposed on the team owing to the previous NVG-solution helmet should be removed and expanded to allow the full potential of the helmet to be tested and utilised. There is no question that any future trial must take place over a wider range of conditions – most definitely as a minimum in the dark! We saw sunset, dusk and a few minutes of darkness but it was clear that the NVC functioned best in darkness and with lower levels of cultural lighting. It was suggested that consideration be given to an opacity filter which would block out the real world scene allowing unambiguous viewing of Zoomed image or other sensor imagery. This would avoid any confusion over multiple images being presented in the visor such as real world, NVC and zoom. With a major positive aspect of the INV being the future potential for other imagery it was felt that this would be worth investigating at least.

Latency was the biggest factor affecting the results of the trial and this was exacerbated by the conditions in which we flew, which created the 'double image' phenomenon. Whilst we did not fly in the ideal conditions we did see the latency issue because of this and therefore we could prioritise a fix. Early flight trial is not always a good thing depending on the criticality of the test item – in this case it was advantageous and we were given early sight of the main issue and identified a need for improvement early. We were able to demonstrate a fix within weeks as there had been an error in the test article that went unnoticed originally.

Partly because we decided to be so ambitious with the profiles, rather than keeping it simple, we ended up performing the trial at the worst possible time of year for such an assessment. ***The main lesson learned, and I have been advised that one is enough, was that when you***

are given objectives, stick with them unless you are prepared to accept the consequences, which may end up being a less efficient trial or much worse. We created a monster and were disadvantaged to a degree by the timing of the trial but were lucky there were no other adverse effects. ***Beware mission creep even if it is not forced upon the test team and seems a good idea at the time.***

It is no surprise that the future will make such technology better and cheaper. Sensor technology advances year on year and we are close to a 2K x 2K resolution NVC becoming available, and whilst this is not a panacea, coupled with increased processing capacity and projection technology INV will become fully effective in all environments, reliable and commonplace. Our next assessment will involve a scan converter to allow the analogue Typhoon aircraft signals to be fully integrated into the digital helmet. We also need to combine the symbology on sensor image assessments at a time of year when we can fly properly in the dark! Colour imagery is already in the pipeline to allow better target and asset recognition and differentiation. As per the RAF Test Pilot's recommendation we will examine other sensors that can be incorporated into the visor display, synthetic vision and NAVFLIR for example. All this technology is great but we always return to the importance of comfort and lighter weight optics will keep the mass down and improve wearability and use in the cockpit. INV is certainly an exciting technology and should make the aircrew task of the operation of aircraft at night significantly safer and more effective, whilst also demonstrating great potential for future development.

LESSONS LEARNED AND MURPHY'S COROLLARY

By Michael Meier

Abstract:
Safety in flight test requires that when the potential consequences of a negative outcome are catastrophic, the go decision must be made only when a very high degree of confidence in a positive outcome exists. When this protocol is followed, it is a natural result that a very high percentage of outcomes will be positive, and it is further likely that we will see a long string of consecutive positive outcomes. When this happens a feedback loop is created in which the go decision is validated by the positive outcome as having been a good decision, and is therefore reinforced and made more likely to be made again in the future under similar circumstances. Beyond that, often the lesson learned from a series of consecutive positive outcomes is that our evaluative criteria for making the go decision were unnecessarily conservative; not only were we correct to make the go decision, but we could have made it under even somewhat less favorable circumstances. This can lead to a gradual lowering of our standards.

However, consecutive successful outcomes can mask go decisions that may be fundamentally unsound, if the success is the result, either partly or entirely, of good fortune. Below a certain threshold of probability for success, repeated go decisions will lead eventually to an unacceptable probability of failure. At 99% probability of success, eighteen events results in a cumulative probability of failure for at least one event of 17%, or one chance in six.

The common idiomatic statement of Murphy's Law is: "Anything that can go wrong will go wrong." This is likely not what Murphy said, nor is it the best statement of the lesson learned from Murphy's experience, and in fact, no one who thinks about it can actually believe it. Almost anything could go wrong, and if it were really true that anything that could go wrong would go wrong, then most everything would go wrong, and most things don't. But if Murphy's Law were true, then a corollary of this "law" would also be true: "Anything that did not go wrong could not have gone wrong." And while we don't really believe Murphy's Law, we are surprisingly susceptible to being seduced by an unconscious belief in Murphy's corollary – that is, we take a positive outcome as evidence that a positive outcome was inevitable.

Combining with our failure to appreciate the nature of probability and our susceptibility to Murphy's corollary is the inevitable and powerful pressure we often experience to make the go decision under less than favorable circumstances. This pressure is a natural and unavoidable consequence of project deadlines, budgets, and performance expectations. It induces us to make the go decision when we otherwise might not, and to validate the go decision afterwards when the outcome is successful, when that validation may be inappropriate.
A means for combating the effects of these three factors is to conduct a rigorous and quantitative pre and post-test analysis not only of failed outcomes, but also of successful ones, with the goal of identifying apparently successful outcomes that may have followed bad decisions, and that were successful because of good fortune.

Background:
The ideas in this paper came out of thinking about a serious hang gliding accident in 1995 that occurred on landing in circumstances under which hundreds of previous successful landings had been made. The ideas were first expressed in an article in Hang Gliding Magazine published in 1998. The article can be read here: https://www.willswing.com/why-cant-we-get-a-handle-on-this-safety-thing/ . They were adapted for presentation to SETP and presented at the 2015 West Coast Symposium. Because the material here was developed for verbal presentation, this paper essentially follows the form of that presentation. Some additional material in the way of amplification and background that could not be included in the presentation due to time constraints is provided. The added material is in italics and indented.

LESSONS LEARNED AND MURPHY'S COROLLARY

Sharing lessons learned, with the goal of enhancing the effectiveness and safety of experimental flight test. That's why we're here. So I want to pose a question: Might we be missing something in this process? I'll suggest an answer a little later on, but right now I'd like to invite you to participate in a thought experiment.

Imagine that you find yourself in need of a significant amount of money, with no obvious way to get it, and with serious negative consequences if you don't. I come along and offer you ten million dollars, if you will play a single game of Russian roulette. One bullet, six chambers, one pull of the trigger. If the gun fires, you're dead. If not, you get the ten million, and your problems are over. OK, I realize your problems would be over in the event of either outcome. Let's disregard that for now. So, the question is, "Would you play?"

I work for a company that designs and manufactures hang gliders.

For 39 years, I've been engaged in the experimental, developmental and production test flying of the gliders we make.
Production test flying is the final quality control check for a manufacturing process that is part science and engineering and part art and craftsmanship, and we've found it to be an essential tool in maintaining manufacturing consistency.

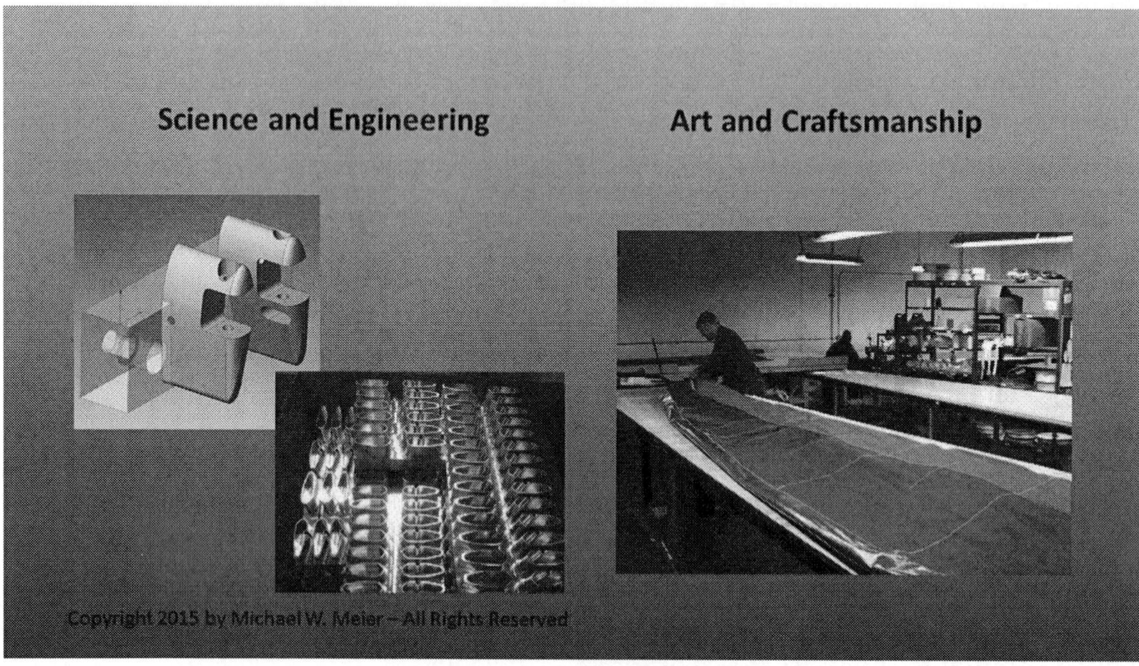

Profit margins in hang gliding are pretty thin, and like most businesses, we continually look for ideas to reduce cost and increase efficiency. About 35 years ago, we hit upon one such idea for production test flying.

Normal procedure involves launching from the top of the mountain, conducting the flight test, and landing in the landing area in the valley below, then driving back up for the next round – a process which might be repeated as many as eight times in a day.

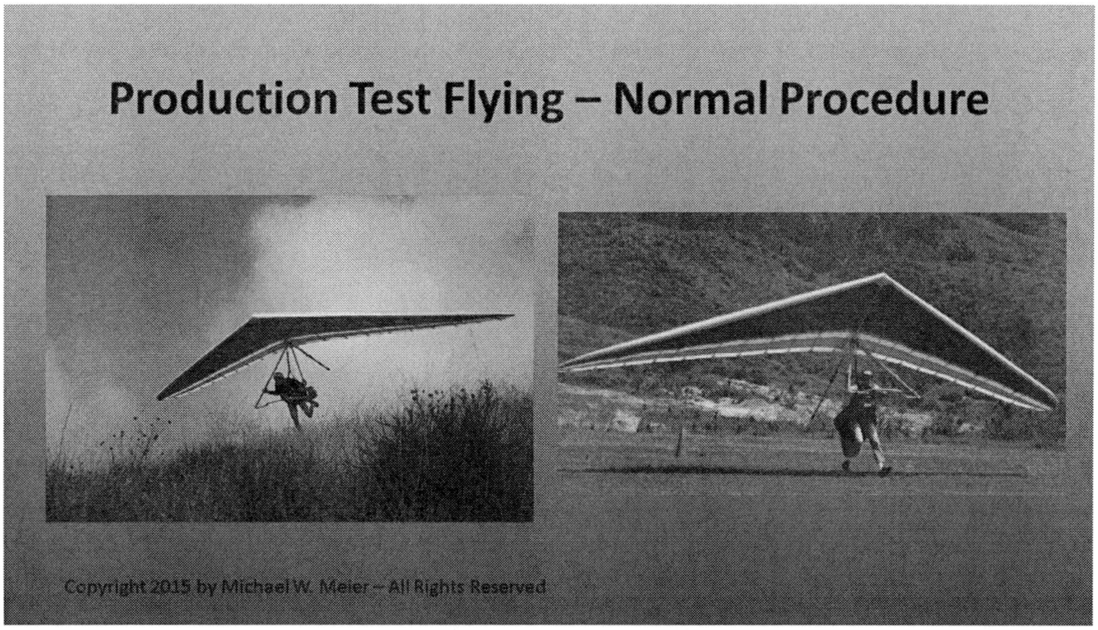

To save time, we thought we could thermal up from launch, perform the flight test, and then land back on top near where we took off from, saving ourselves the drive back up from the landing area and eliminating the need for one crew member - the driver who would otherwise be needed to drive the truck down the mountain to pick us up after each round.

Not every hang gliding launch site lends itself to top landing, but one of ours did.

Still, the landing demands precision and the right approach.

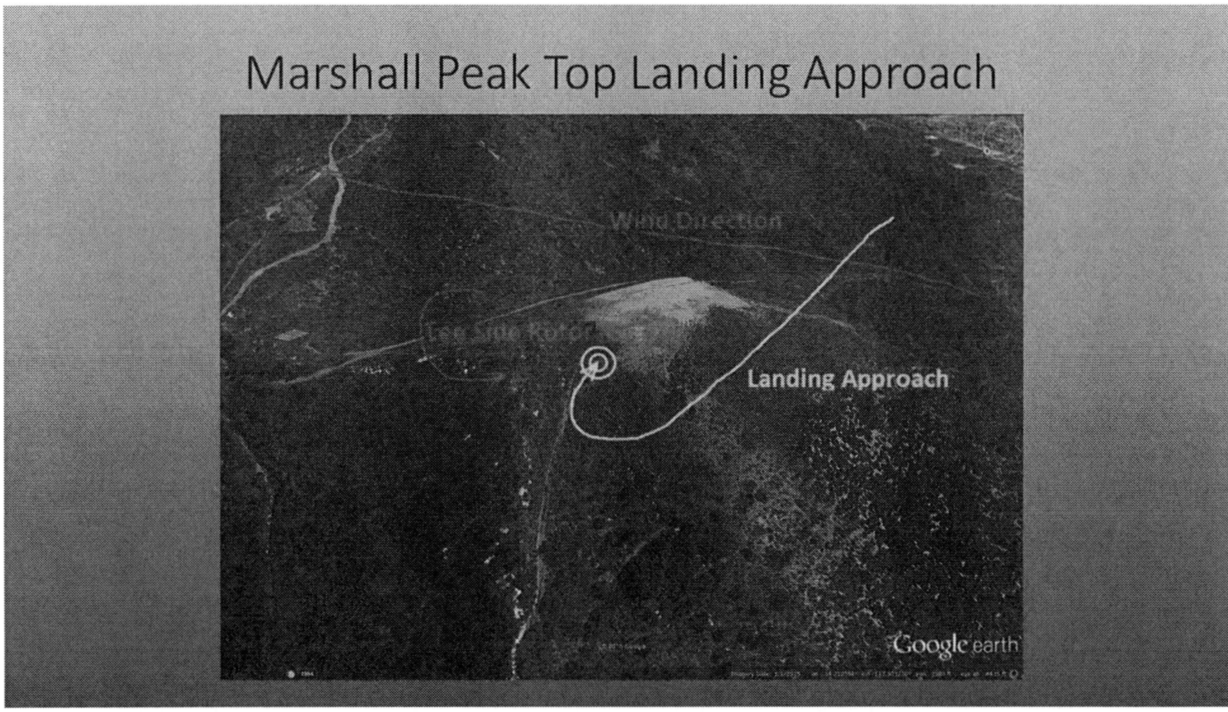

Try to approach into the wind and there are two likely outcomes – either you'll be too high or too fast and overshoot, or you'll be too low or too slow, and descend into the lee side rotor behind and below the hill and lose control in the turbulence. The window of "just right" between the two is extremely small.

So we developed an alternate approach – coming in half crosswind, more from the side of the hill than from the back, approaching with good speed and landing on the upslope, slightly below the top.

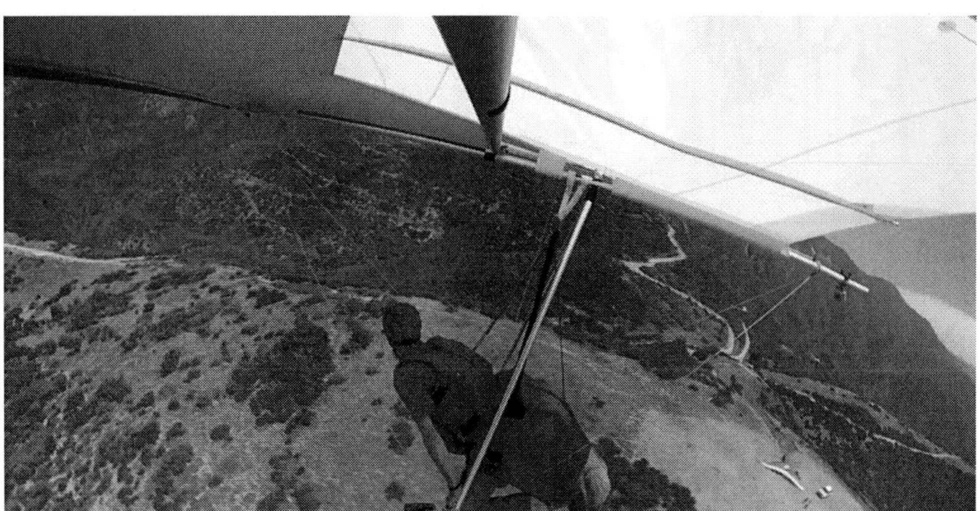

"Downwind" leg

Continuous turn through base to final

Flaring for landing on the upslope, below the hill top

And we did this for more than fifteen years, when conditions allowed for it – and in that time I probably had fifteen hundred or more successful such landings. And then one day I didn't: I lost control of the glider in a strong piece of turbulence just before landing flare, and I crashed hard.

That was not my crash landing, but that's about what mine would have looked like. For a brief moment after impact I thought I might be dead. For a moment longer, I thought I might be paralyzed. In the end, I got away with a sprained ankle and a moderate case of whiplash.

The accident had happened just prior to a family vacation abroad, and so I had three weeks to think about it while we bounced around on the rutted dirt roads of East Africa and I tried to hold my head still to mitigate the pain in my neck.

As I tried to figure out what had gone wrong, I went over every aspect of the approach. I was flying an intermediate model glider - not very demanding of pilot skill. The conditions were typical for summer – thermals and active air, but not particularly strong or turbulent. I had done hundreds of landings in stronger conditions, on more demanding gliders. At the same time, I was neither sloppy nor complacent in my approach. Though I was relaxed, I was also focused. My intent was simply to fly a perfect approach, and I knew exactly where I wanted to be at every point during the approach – position, heading, altitude and airspeed. On that day I executed the approach exactly as I wanted to.

Finally, after a lengthy analysis during which I could uncover no mistakes in my execution, I was left with only one conclusion. Given the result, which with a little less luck could easily have been a serious injury or death, that landing attempt had been inherently, significantly dangerous.

Consequently, it followed that the error I had made was in the decision to attempt the landing. That decision had been a bad decision, proven so by what had happened, and by what, with a little less luck, could have happened. And given that most of the top landings I had done over the previous fifteen years had involved gliders and conditions that were as demanding or more demanding, that meant that every one of those other decisions had also been bad decisions. And, they had been bad decisions in spite of the fact that no bad outcomes had occurred as a result of any of them. And that's when the thought occurred to me that I was understanding something for the first time.

My talk today is based on two fundamental ideas: First that the primary determinant of safety is the quality of decision making, and second, that safety in flight test requires that when a negative outcome has potentially catastrophic results, the go decision must be made only when a very high degree of confidence in a positive outcome exists.

Now when this protocol is followed, it is a natural result that a very high percentage of outcomes will be positive, and it is therefore also likely that we will see a long string of consecutive positive outcomes. When this happens, a feedback loop is created, in which each go decision is validated by the positive outcome as having been a good decision, and therefore is reinforced and becomes more likely to be made again in the future under similar circumstances.

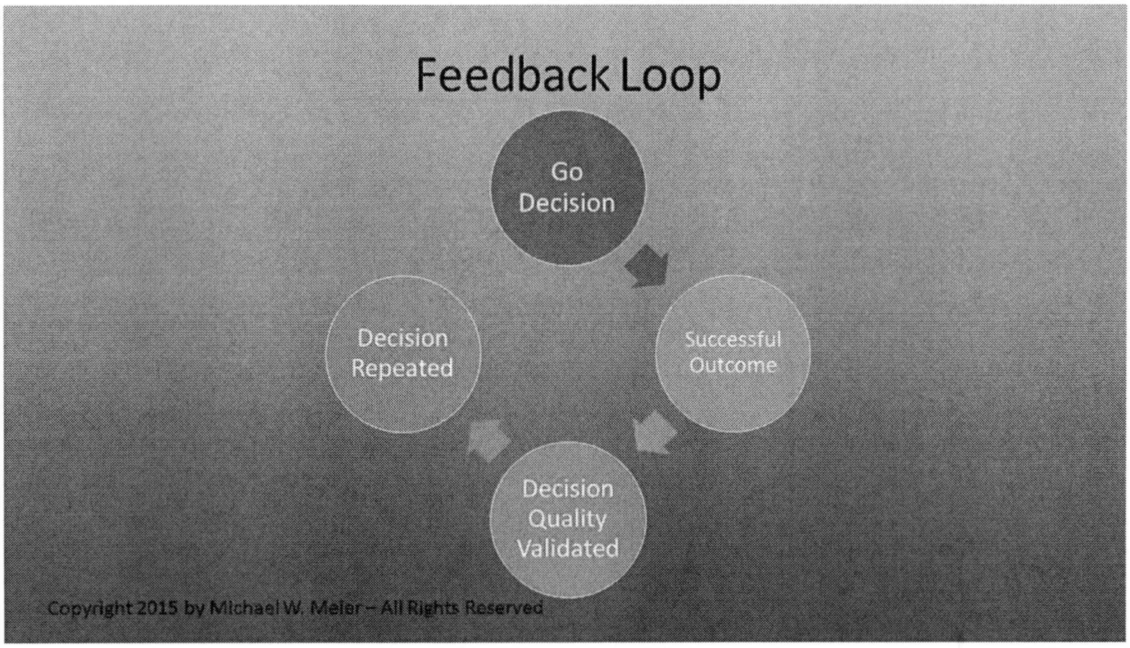

But it goes beyond that.

Often, the "lesson learned" from a string of successful outcomes, even if unconsciously, is that our evaluative criteria for the decisions were unnecessarily conservative – not only were we correct to make the go decisions, but we had margin enough that we could have made them even under somewhat less favorable conditions. This can lead to a gradual lowering of our standards.

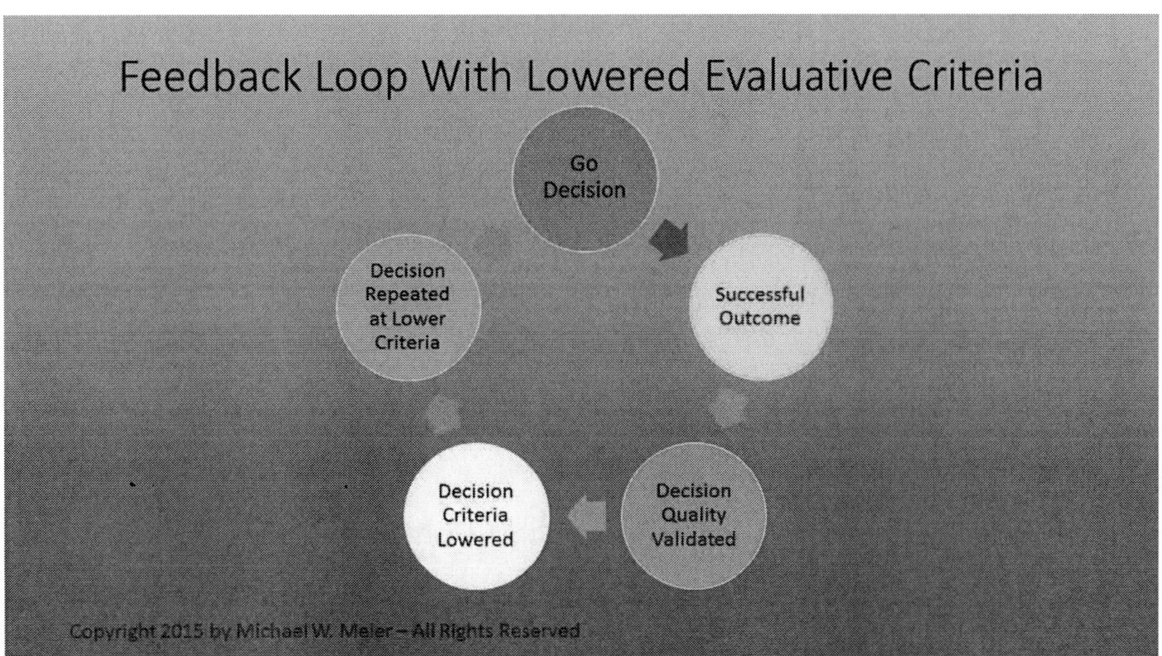

I want to talk about Murphy's Law, or at least the statement of Murphy's law that has gone into the popular lexicon – "Anything That Can Go Wrong Will Go Wrong." That's probably not what Murphy said, but that's how we quote him today. And when we quote Murphy in this way, it is likely that we are doing it mostly for comic or ironic effect. Because to be clear, nobody really believes this statement of Murphy's Law. After all, almost anything could go wrong, and if it were really true that anything that could go wrong would go wrong, then most everything would go wrong. And most things don't.

> *The origin of Murphy's law is generally attributed to a statement allegedly made by Capt. Edward A. Murphy, a development engineer from Wright Field Aircraft Lab working on Air Force Project MX981 at Edwards Air Force Base in 1949. Commonly known as the rocket sled tests, the project was designed to see how much sudden deceleration a person could withstand in a crash. Murphy had brought a G force sensor, consisting of a bridge of four strain gauges. The gauges could be wired together in more than one way, and in one of those possible ways would cancel each other out and produce a zero reading. After just such a result, Murphy allegedly remarked in frustration, "If there is any way to do it wrong, he will" – referring to the technician who had wired the gauges. This has since been generalized to the now common statement of Murphy's Law – Anything That Can Go Wrong Will Go Wrong.[1,2] This basic observation can be found at least 70 years before Murphy's statement, and likely goes back much further than that, but today it is almost universally attributed to Murphy.[3] The statement is obviously false, since almost anything could go wrong and most things don't, but it is only false because the common phrasing omits one important word and the related concept. What the statement of the law should be is, "anything that can go wrong, will go wrong, eventually." That is to say, given enough trials, if something can go wrong, it eventually will. The value of Murphy's observation is not in the relatively pessimistic implication of the Law as it is quoted today, but rather in the more meaningful way in which it informs design – both the design of things, and the design of procedures. What Murphy is really saying is that if we can design out the possibility of doing it wrong, we should. Our lives today are full of examples of this – the polarized electrical plug being just one.*
>
> *A tragic illustration of Murphy's Law was the series of crashes in the F-86's in which a bolt on the aileron cylinder was installed upside down causing the ailerons to lock up under certain loading conditions. The incorrect bolt installations were reportedly traced to one older worker on the assembly line, who ignored the instructions regarding the orientation of the bolt because he felt he knew better how the bolt should be installed.[4]*

But if Murphy's law were true, if it were really true that anything that could go wrong would go wrong, then a corollary of Murphy's law would also be true: "Anything That Did Not Go Wrong, could not have gone wrong."

And, interestingly enough, even though we don't believe Murphy's Law, we are surprisingly subject to being seduced by this corollary of Murphy's Law - that is, if nothing DID go wrong, then nothing COULD HAVE gone wrong. In other words, we take a positive outcome as evidence that a positive outcome was inevitable.

This is what happened with our decision to use top landings to improve efficiency. We had a long string of successes, each one of which reinforced our perception that success was an outcome we could rely on. And here's the problem with that perception - our data set is too small, at least in the beginning of any such process.

> *It's important to an understanding of the power of Murphy's corollary to note that we had many indications during those fifteen years of successful landings that the decision to use this technique was of questionable safety. There were a couple of reasons we chose to land on the upslope of the side of the hill, below the top. One reason was that there was inevitably some turbulence, either rotor or convective or both, as we passed through an altitude of about twenty to fifty feet agl. Being weight shift controlled aircraft with relatively large wingspans and light wing loading (33 foot span, 1.1 to 2.0 lbs per square foot) hang gliders suffer from a progressive loss of control authority and response at lower speeds, especially as one approaches stall speed while slowing down. Approaching from below the top allowed us to approach at high speed for better control, and landing on the upslope greatly reduce the duration of time spent at low speed, reducing the probability of a loss of control. Typically, we would pass through the area of turbulence while still flying fast, then enter a region of wind gradient, where airspeed would drop dramatically as we rounded out to follow the slope of the hill, and then almost immediately flared for landing. Many times, while passing through the region of turbulence and then entering the gradient, I remember thinking, "this may not end well." And yet, with each successful landing, my next thought was, "well that was fine, no problem – what could go wrong?"*

> *I had even observed other pilots having significant crash landings, sometimes even damaging gliders, yet I was able to rationalize those as having been due to poor technique – a poorly executed approach path, or poor control of airspeed on approach. It was only when it happened to me, following what I deemed to be a perfectly executed approach, that I was able to realize that my doubts about the safety had in fact been correct, and the suppression of those doubts had been Murphy's corollary in action.*

Let's imagine we're making "Go" decisions when we have a 99 per cent probability of success. Chances are, we'll see a long string of successful outcomes. And yet, after only 18 such decisions, our overall cumulative probability of success has fallen to 83%.

> *Raise the percentage probability of an outcome to the power of the number of trials, to compute the overall percentage probability of that outcome for all events in the series.*

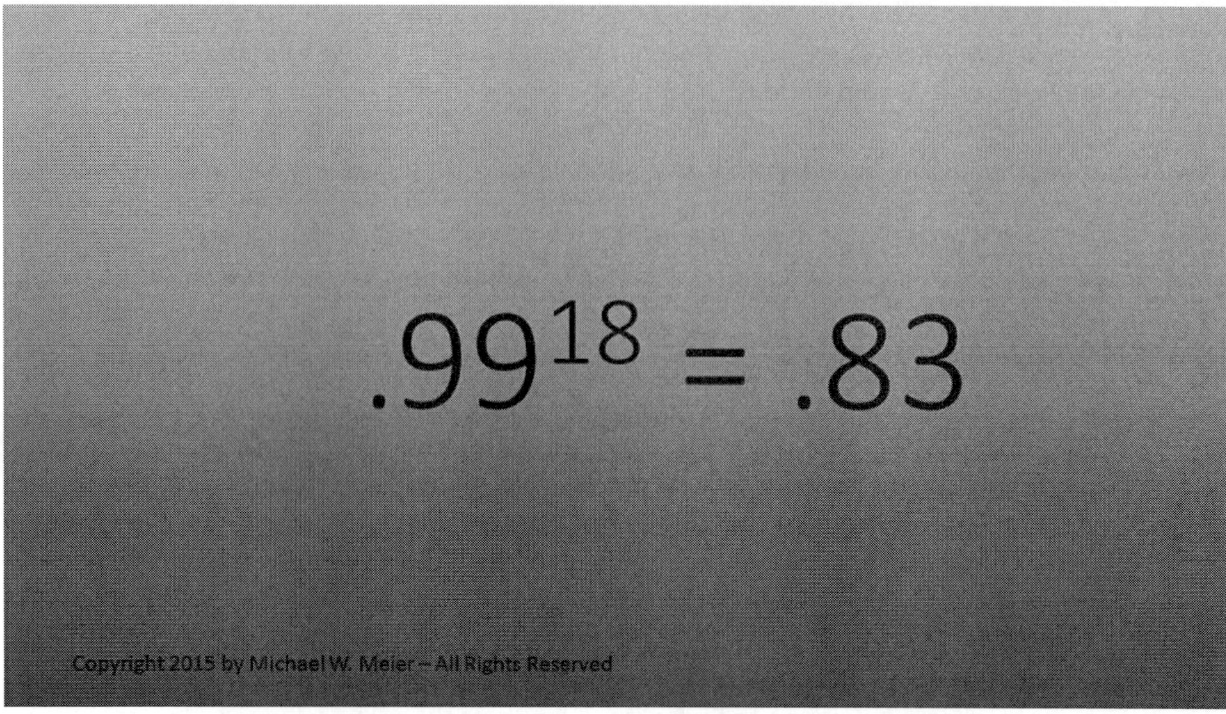

In other words, after 18 events, our cumulative probability of failure has risen to 17%, or one in six - the same level as in our game of Russian roulette.

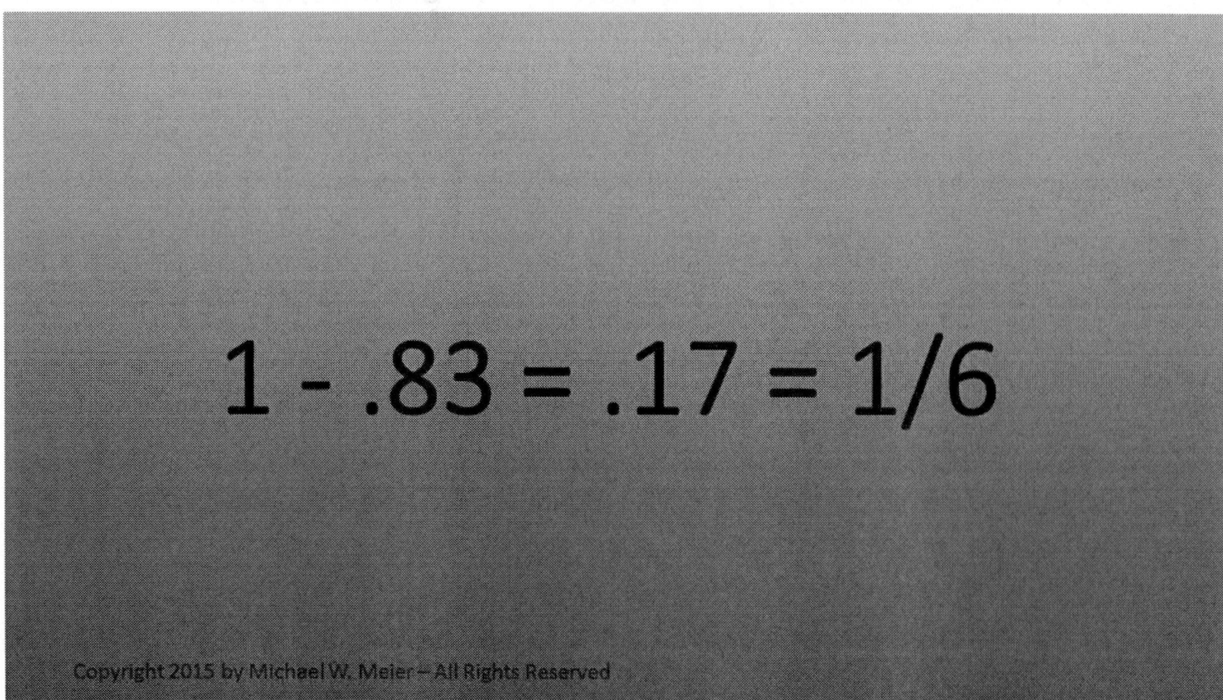

I don't know how many of you took me up on my offer of ten million dollars with one chance in six of a catastrophic outcome, but my guess is not many.

And finally, if all of the foregoing were not enough, there is one very significant additional factor that comes into play – the inevitable pressure on us to make the go decision.

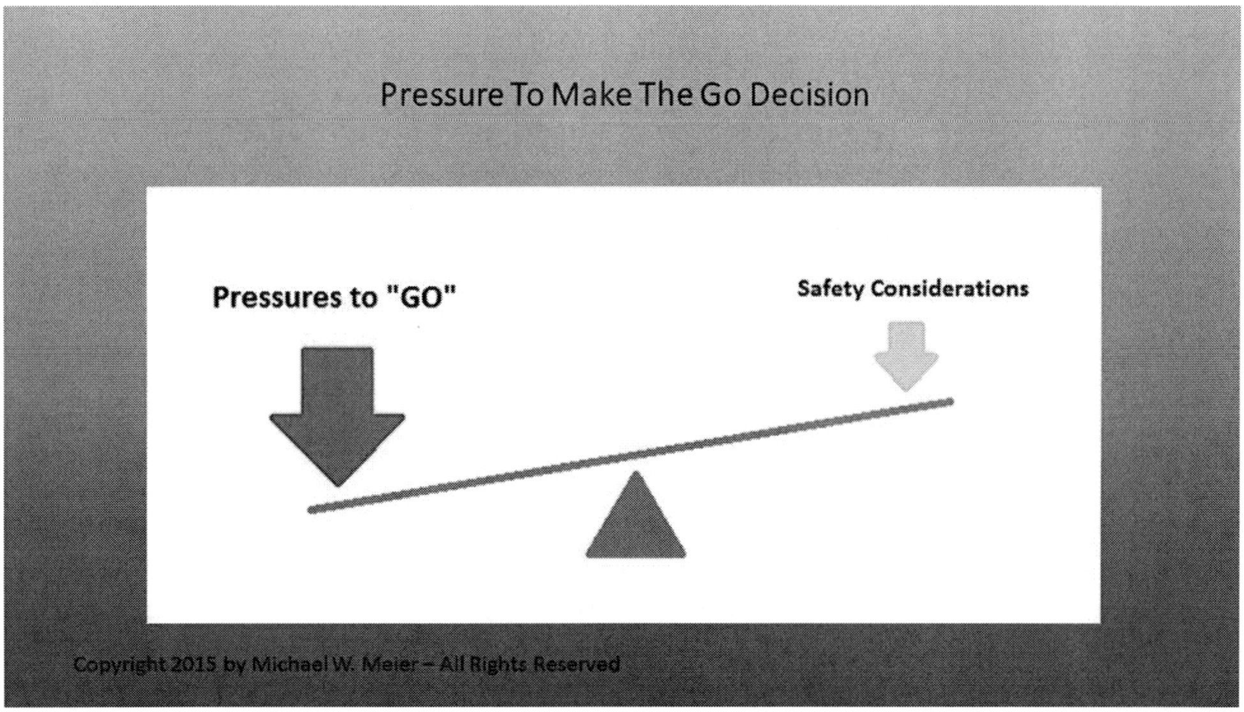

This pressure can take many forms, and have many sources, an impending deadline, a project that is behind schedule or over budget or below expectations, or simply our own personal ego, but it is almost always present in some form. The result is that we have a strong incentive to make the go decision, even when we have some level of doubt about it. And beyond that, we have a strengthened tendency to validate the go decision as having been a good one when the outcome is successful, because now the pressure has been relieved.

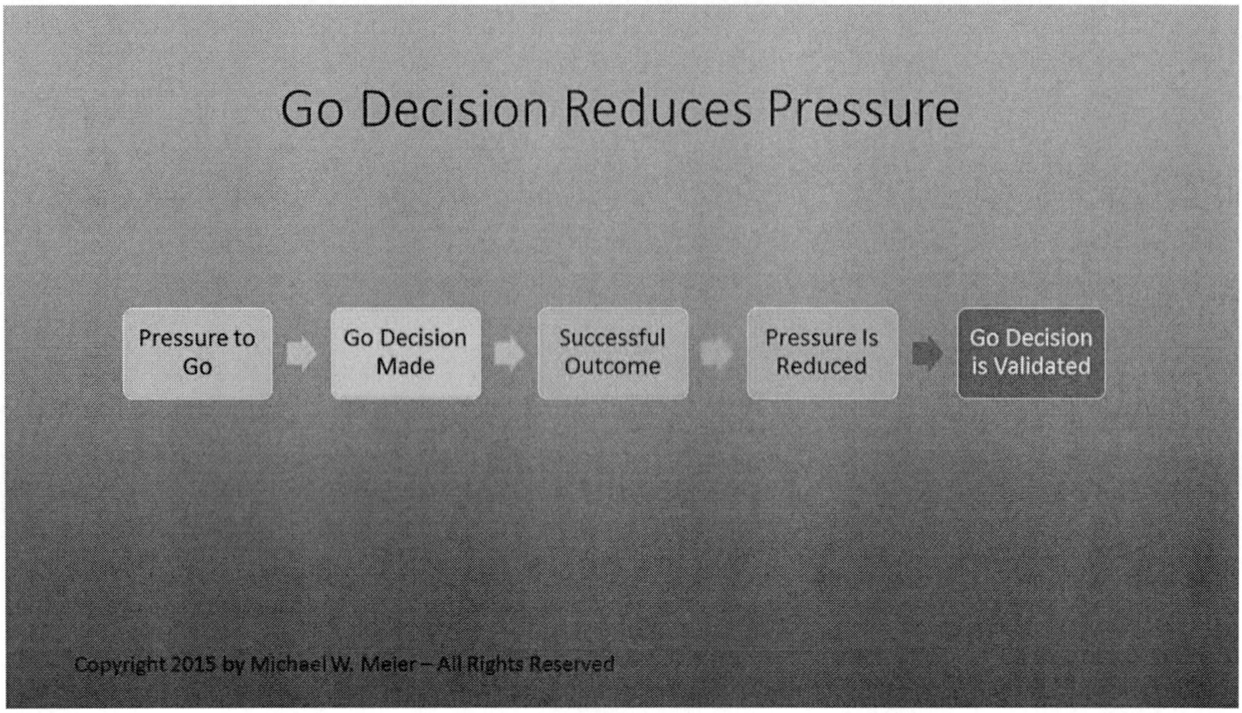

Conversely, if we do decide to make the no go decision, then the pressure to make the go decision the next time typically only increases, as we are now farther behind on whatever schedule, or budget, or whatever other factors may have been creating the pressure to go in the first place.

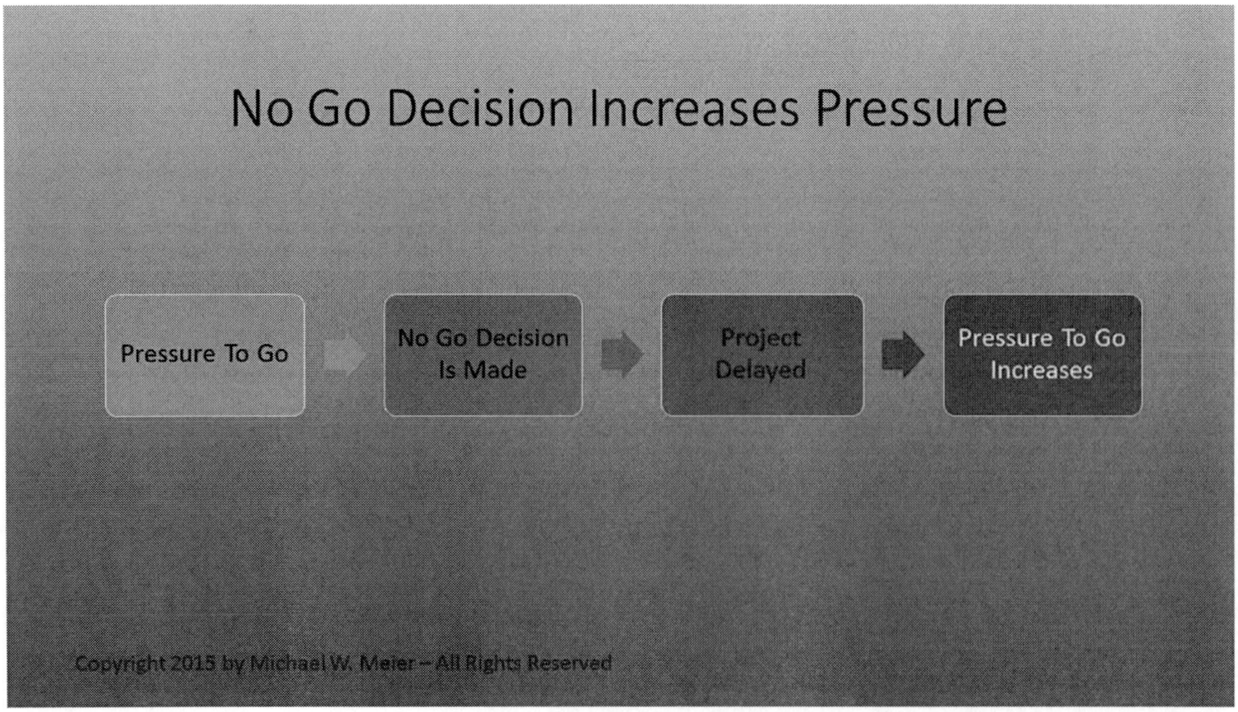

> *I have, over time, come to the conclusion that the pressure to go may be the single most difficult factor to overcome in the pursuit of safety. I have observed in my own decision making that I am far from immune to it, despite the extent of the analysis I have done into its potential for negatively impacting the decision making process in relation to safety.*

Ok. So twenty years ago I had a hang gliding accident that led me to start thinking about how we make decisions that impact our safety. Three years after that, I wrote these thoughts out in some detail, in much the same form as I've talked about them today, in an article for the national Hang Gliding Magazine. And eventually I began to ask myself whether any of these ideas might have some broader relevance. And you might be asking the same question at this point - how does any of this apply to experimental flight test in the field of aerospace?

I will suggest three possible examples:

We'd had a history of success with a pressurized pure oxygen environment in manned space capsules, until Apollo One. Had we thought sufficiently critically about it, we might have concluded that a complex electrical system in a significantly pressurized pure oxygen environment was a dangerous condition. We had even had previous fires in such environments, and we might have drawn on those as evidence of such danger.

But, in the words of Frank Borman, the astronaut's representative on the Apollo 204 Review Board, in testimony before the House Subcommittee on NASA Oversight,

"I don't believe that any of us recognized that the test conditions for this test were hazardous."[5] "We did not think, and this is a failing on my part and on everyone associated with us; we did not recognize the fact that we had the three essentials, an ignition source, extensive fuel and, of course, we knew we had the oxygen."[6]

In the Summary of its report, the Committee on Aeronautical and Space Sciences of the United States Senate had this to say:

"It is clear from the Board's report and the testimony before the committee that this kind of accident was completely unexpected despite the amount of documentation of fire hazards in pure oxygen environments. The committee can only conclude that NASA's long history of successes in testing and launching space vehicles with pure oxygen environments at 16.7 p.s.i. and lower pressures led to overconfidence and complacency."[7]

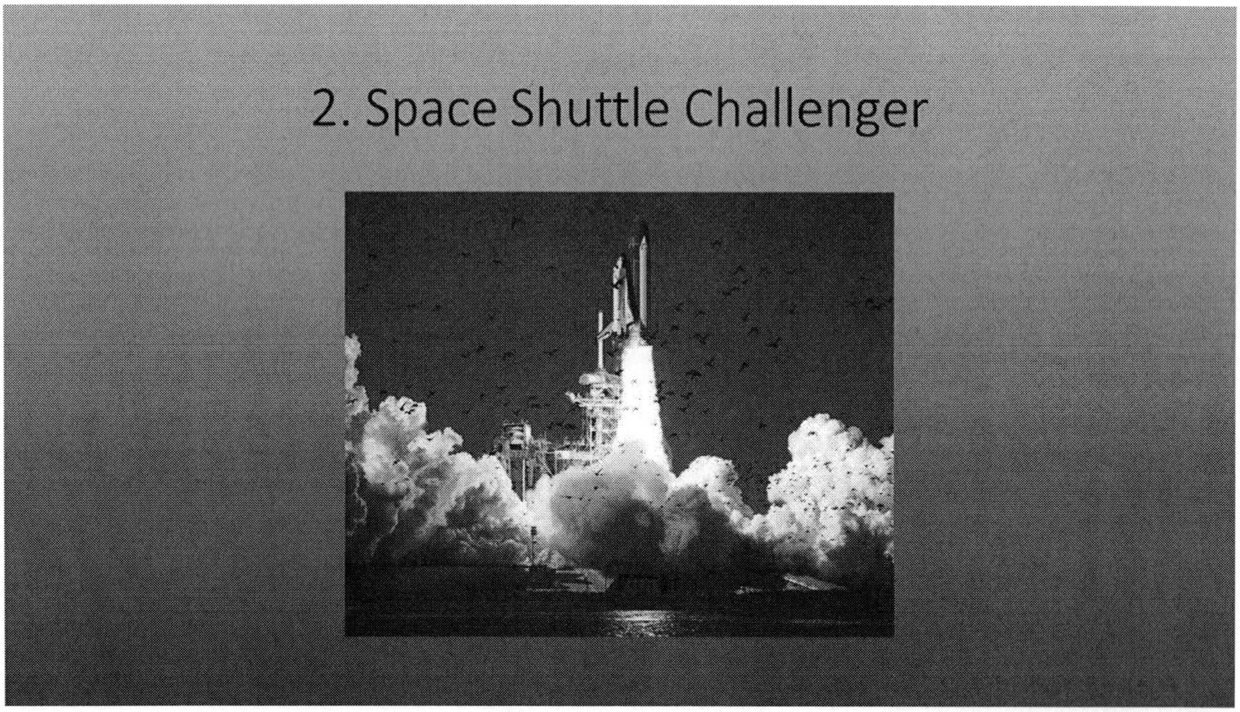

We had successfully launched a number of space shuttles with O- ring erosion and hot gas blow by in the joints of the solid rocket boosters adjacent to the external tank. With sufficiently critical analysis, and better internal communication, we might have concluded that this was a dangerous situation before we had one blow up. Instead, influenced by our repeated successes, we actually expanded both the amount of blow by that was acceptable and the low temperature range at which we were willing to launch, increasing the danger of a catastrophic failure.

Quoting from the Report of the Presidential Commission on the Space Shuttle Challenger Accident:

"The Commission concluded that there was a serious flaw in the decision making process leading up to the launch of flight 51-L. A well-structured and managed system emphasizing safety would have flagged the rising doubts about the Solid Rocket Booster joint seal."[8]

"Neither Thiokol nor NASA expected the rubber O-rings sealing the joints to be touched by hot gases of motor ignition, much less to be partially burned. However, as tests and then flights confirmed damage to the sealing rings, the reaction by both NASA and Thiokol was to increase the amount of damage considered "acceptable." At no time did management either recommend a redesign of the joint or call for the Shuttle's grounding until the problem was solved."[9]

"NASA and Thiokol accepted escalating risk apparently because they 'got away with it last time.' As Commissioner Feynman observed, the decision making was: 'a kind of Russian roulette. ... (The Shuttle) flies (with O-ring erosion) and nothing happens. Then it is suggested, therefore, that the risk is no longer so high for the next flights. We can lower our standards a little bit because we got away with it last time. ... You got away with it, but it shouldn't be done over and over again like that.'"[10]

We had successfully re-entered and landed a number of shuttles with damaged or missing heat shield tiles that had been struck by foam shedding from the external tank. We might have concluded that the damage to or loss of those tiles created an unacceptable risk on re-entry.

Instead, we continued to launch shuttles with the foam shedding from the tank and damaging the tiles, until we lost Columbia.

Quoting from the Report Of The Columbia Accident Investigation Board:

"The shedding of External Tank foam – the physical cause of the Columbia accident – had a long history. This raises an obvious question: Why did NASA continue flying the Shuttle with a known problem that violated design requirements? It would seem that the longer the Shuttle Program allowed debris to continue striking the Orbiters, the more opportunity existed to detect the serious threat it posed. But this is not what happened"[11]
"With each successful landing, it appears that NASA engineers and managers increasingly regarded the foam-shedding as inevitable, and as either unlikely to jeopardize safety or simply an acceptable risk."[12]

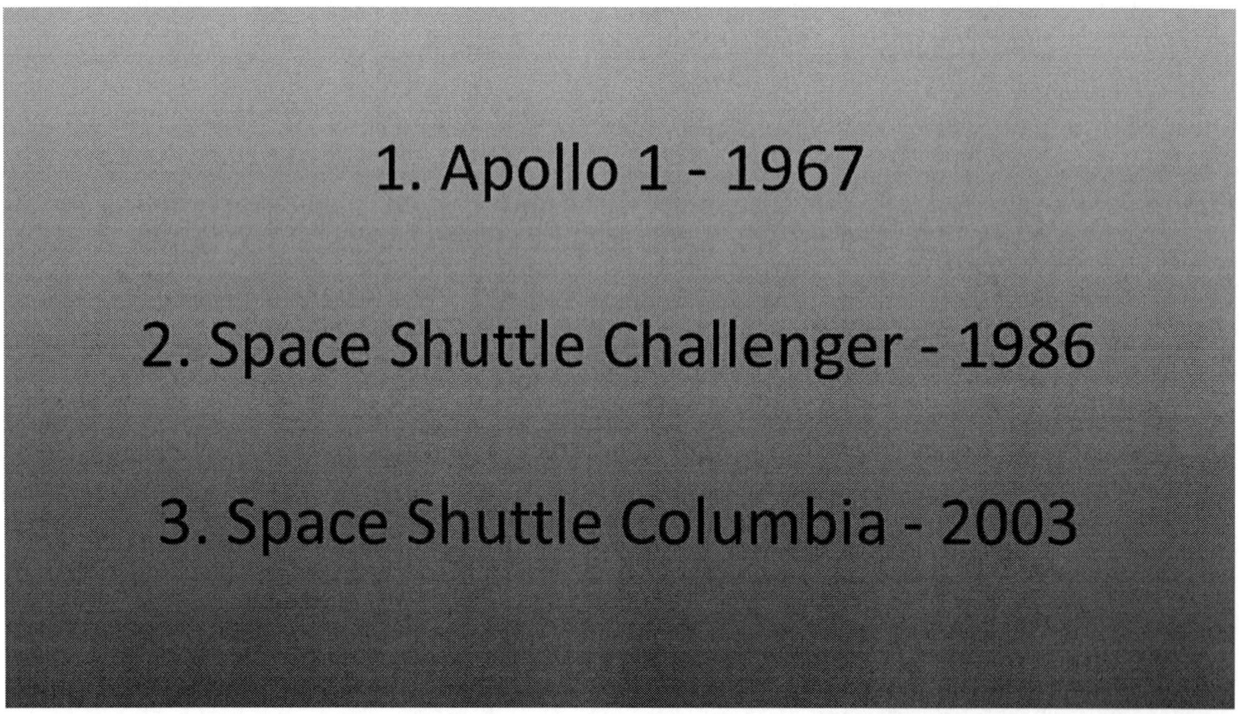

In each of these three cases, the post-accident investigation found that a problem had existed that should have been recognized and corrected, and was not.

I've offered these incidents as evidence of relevance to the Society of the ideas I've talked about today. I've quoted from several accident reports and in so doing, I have clearly demonstrated that nothing I have said here today is either new or original, even though these ideas seemed both new and original when they first occurred to me, as they came out of my analysis of my own accident, and as I had not read any of these NASA reports at that time.

What may be somewhat original, or at least worthy of consideration, in what I've talked about today, are some possible insights into why these problems seem so intractable. Why, after these factors have been repeatedly identified in formal accident investigation reports, do they then show up again in subsequent accidents? Why did we have Challenger after Apollo One, and why did we have Columbia after Challenger?

The Report Of The Columbia Accident Investigation Board, among other recommendations, called for a "new culture" at NASA.[13]

I will suggest that as decision makers we can become victims not merely of a culture, but of a deeper and more persistent human psychology.

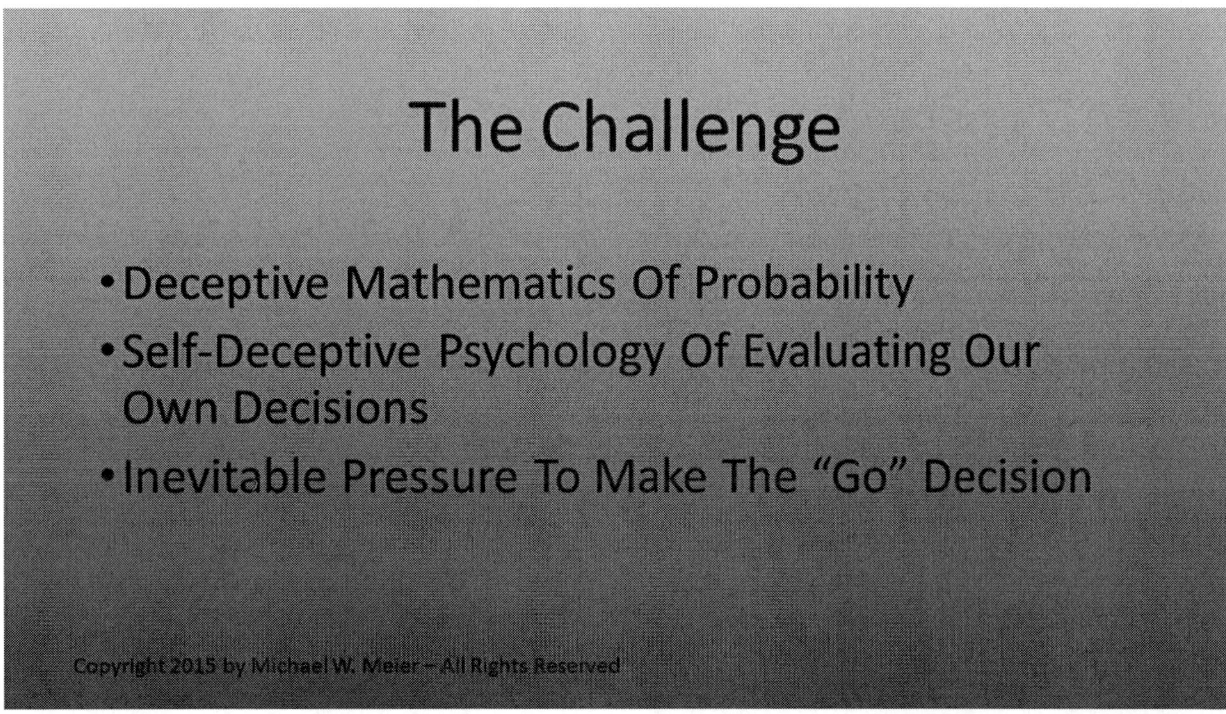

I think that as decision makers we face an unexpectedly difficult challenge arising out of a powerful combination of the inherently deceptive mathematics of probability, the potentially self-deceptive psychology involved in evaluating the quality of our own decisions, and the inevitable significant pressure we experience to make a go decision even when we have doubts about the safety of that decision.

So, if this is a problem, then what is the answer?

I won't suggest that I have an answer, but I will offer two strategies that I have employed, and that I have found to be helpful.

The first is simply to be aware of the problem; to be continually aware of our inherent tendency to succumb to Murphy's Corollary, and to the pressure to go forward under less than adequately favorable circumstances.

The second is to expand the scope of source material for our analysis into lessons learned.

We tend to share lessons that are learned from some type of flight test failure. And we analyze these failures, and share these lessons in the hope and belief that doing so will help us avoid repeating the mistakes of the past. And this is of course completely appropriate and worthwhile.

At the outset, I asked if this process might be missing something. This is what I think we may be missing - that there may also be techniques available for identifying and avoiding mistakes of the future – mistakes we don't know anything about because we've not yet seen them.

This might be achieved through a careful and critical pre and post-test analysis of our successes – an analysis aimed at identifying successes which may be wholly, or partially, the result of good fortune as opposed to sound decision making.

One method for doing this is to establish beforehand a quantitative metric that relates to safety, and then to set a target for performance. The target can be set so as to allow for a significant margin of safety.

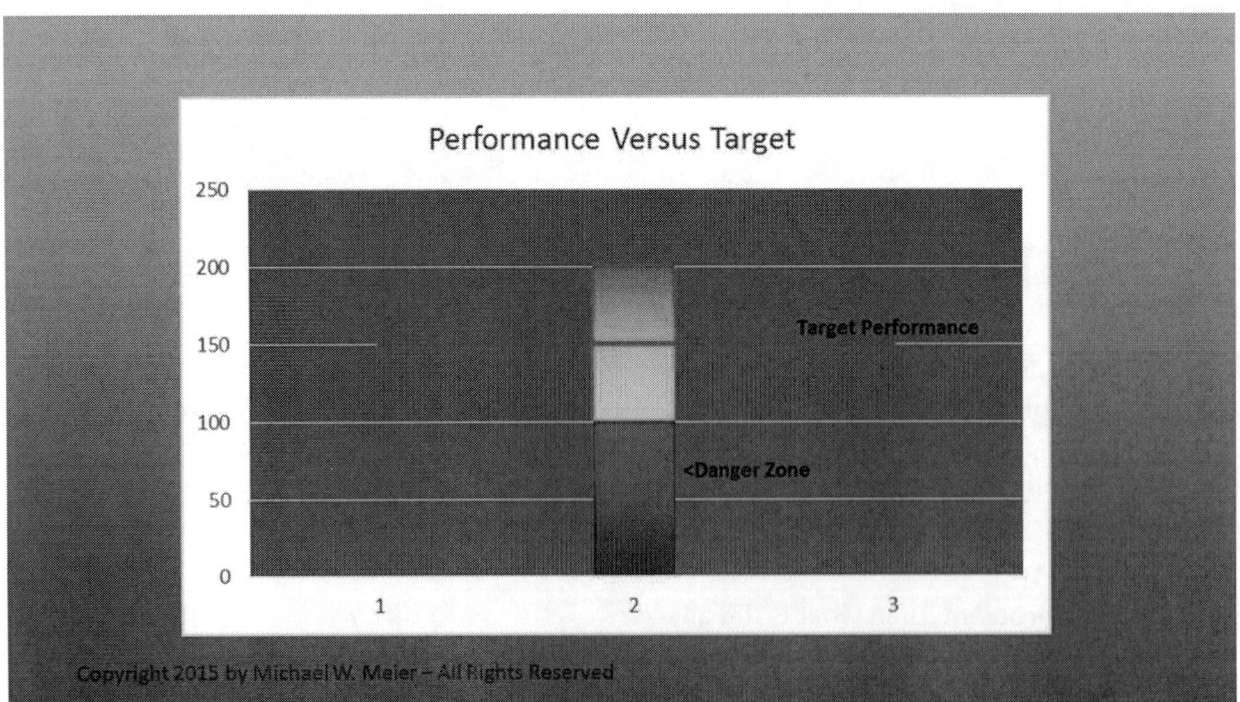

Then by measuring actual performance against our target, we may identify apparently successful outcomes that did not meet our target performance, and therefore should not, perhaps, be considered successes.

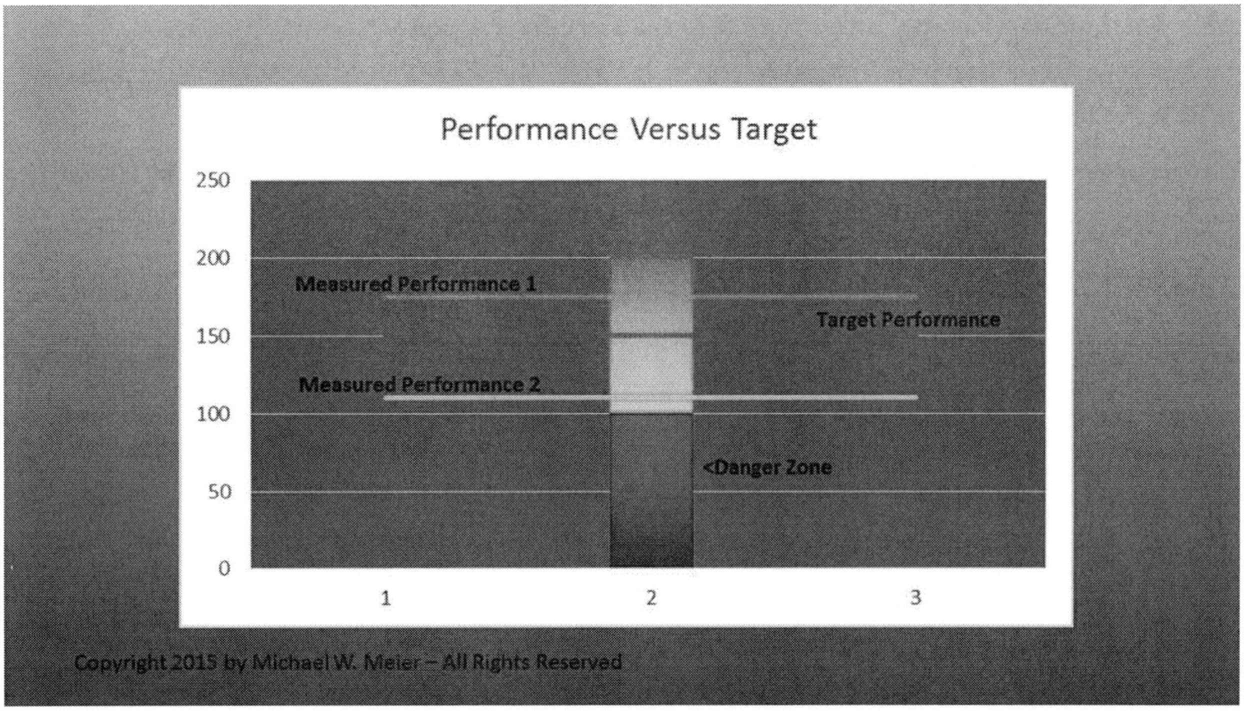

This could have been done during the Shuttle program in relation to the O-ring seals, and it would have triggered a stop to the program until the problem was fixed. We might guess why that wasn't done – the program was under performing and the pressure to go, which can exist at many levels within a complex organization, was likely significant.

> *The existence and / or perception of pressure was referenced a number of times in the Rogers report. The commission noted for example: "The nation's reliance on the Shuttle as its principal space launch capability created a relentless pressure on NASA to increase the flight rate."*[14] *Roger Boisjoly, a structures engineer at Thiokol, who strongly opposed the launch of Challenger under the temperature conditions at the time, testified before the commission, "I felt personally that management was under a lot of pressure to launch and that they made a very tough decision, but I didn't agree with it."*[15] *Interestingly, it does not really matter whether pressure is "real" or merely perceived – it has the same effect. In Moonshot, Alan Shepard and Deke Slayton wrote, regarding Skip Chauvin's decision to decline an engineer's suggestion to cancel the plugs out test due to multiple problems, "Time was more important. It was becoming damned critical with all the pressures mounting to get this ship on its way into orbit."*[16]

Ironically, however, it's not unlikely that the program lost more time in fixing the problem after the catastrophe than would have been lost to a proactive approach to fixing it ahead of time.

A rigorous analysis of successful outcomes would constitute an expansion of our source material for lessons learned. If such a process allows us to uncover some decisions or procedures that may have been in some way flawed, we may expand the degree to which our lessons learned enhance flight test safety.

Acknowledgments

The author gratefully acknowledges the assistance of Colin A. Fries and the NASA HQ History Program Office in gaining access to original source material used for this paper.

Notes:

1. "Murphy's laws origin - Excerpted from The Desert Wings, March 3 1978" Murphy's laws site Accessed September 5, 2015 http://www.murphys-laws.com/murphy/murphy-true.html

2. Spark, Nick T. The Road To Murphy's Law Improbable Research 2003 http://www.improb.com/airchives/paperair/volume9/v9i5/murphy/murphy1.html

3. "Murphy's Law" Wikipedia Accessed September 5, 2015 https://en.wikipedia.org/wiki/Murphy%27s_law

4. Yeager, Chuck and Leo Janos *Yeager, An Autobiography* New York: Bantam Books, 1985 185

5. Investigation Into Apollo 204 Accident - Hearings Before The Subcommittee On Nasa Oversight of the Committee on Science And Astronautics U.S. House Of Representatives Ninetieth Congress April 10, 1967 Evening Session *history.nasa.gov* Accessed September 5, 2015 http://history.nasa.gov/Apollo204/bormanhouse1.pdf, 81

6. Ibid., 87

7. Anderson, Clinton P. et al "Apollo 204 Accident - Report of the Committee On Aeronautical and Space Sciences - United States Senate - Report N. 956" *NASA Historical Reference Collection, NASA History Office, NASA. Headquarters, Washington, DC* January 30, 1968 http://history.nasa.gov/as204_senate_956.pdf 9, 10

8. Rogers, William P. et al "Report of the PRESIDENTIAL COMMISSION on the Space Shuttle Challenger Accident" *NASA Historical Reference Collection, NASA History Office, NASA Headquarters, Washington, DC* June 6, 1986 http://history.nasa.gov/rogersrep/genindex.htm Chapter V – Findings [104]

9. Idid., Chapter VI – Early Design

10. Ibid., Chapter VI – Findings, Item 3.

11. Gehman, Jr., Harold W. et al "Report of Columbia Accident Investigation Board" *National Aeronautics and Space Administration* August 26, 2003 http://www.nasa.gov/columbia/home/CAIB_Vol1.html Chapter 6.1

12. Ibid.

13. Ibid. Chapter 11 – Recommendations

14. Rogers, et al, op. cit., Chapter IX, Section VII – Flight Rate

15. Ibid., Chapter V

16. Shepard, Alan and Deke Slayton *Moonshot* Atlanta: Turner Publishing, Inc., 1994, 198

References:

1. Anderson, Clinton P. and Richard B. Russell and Warren G. Magnuson and Stuart Symington and John Stennis and Stephen M. Young and Thomas J. Dodd and Howard W. Cannon and Spessard L. Holland and Walter F. Mondale and Margaret Chase Smith and Bourke B. Hickenlooper and Carl T. Curtis and Len B. Jordan and Edward W. Brooke and Charles H. Percy "Apollo 204 Accident - Report of the Committee On Aeronautical and Space Sciences - United States Senate - Report N. 956" *NASA Historical Reference Collection, NASA History Office, NASA. Headquarters, Washington, DC* January 30, 1968 http://history.nasa.gov/as204_senate_956.pdf

2. Gehman, Jr., Harold W. and John L. Barry and Duane W. Deal and James N. Hallock, Ph.D. and Kenneth W. Hess and G. Scott Hubbard, and John M. Logsdon, Ph.D., and Douglas D. Osheroff, Ph.D. and Sally K. Ride, Ph.D. and Roger E. Tetrault and Stephen A. Turcotte and Steven B. Wallace and Sheila E. Widnall, Ph.D. "Report of Columbia Accident Investigation Board" *National Aeronautics and Space Administration* August 26, 2003 http://www.nasa.gov/columbia/home/CAIB_Vol1.html

3. Rogers, William P. and Neil A. Armstrong and David C. Acheson and Dr. Eugene E. Covert and Dr. Richard P. Feynman and Robert B. Hotz and Major General Donald J. Kutyna and Dr. Sally K. Ride and Robert W. Rummel and Joseph F. Sutter and Dr. Arthur B. C. Walker, Jr. and Dr. Albert D. Wheelon and Brigadier General Charles Yeager and Dr. Alton G. Keel, Jr. "Report of the PRESIDENTIAL COMMISSION on the Space Shuttle Challenger Accident" *NASA Historical Reference Collection, NASA History Office, NASA Headquarters, Washington, DC* June 6, 1986 http://history.nasa.gov/rogersrep/genindex.htm

4. Shepard, Alan and Deke Slayton *Moonshot* Atlanta: Turner Publishing, Inc., 1994

5. Spark, Nick T. The Road To Murphy's Law *Improbable Research* 2003 http://www.improb.com/airchives/paperair/volume9/v9i5/murphy/murphy1.html

6. Yeager, Chuck and Leo Janos *Yeager, An Autobiography* New York: Bantam Books, 1985

7. Investigation Into Apollo 204 Accident - Hearings Before The Subcommittee On Nasa Oversight of the Committee on Science And Astronautics U.S. House Of Representatives Ninetieth Congress April 10, 1967 Evening Session *history.nasa.gov* Accessed September 5, 2015 http://history.nasa.gov/Apollo204/bormanhouse1.pdf

8. "Murphy's Law" *Wikipedia* Accessed September 5, 2015 https://en.wikipedia.org/wiki/Murphy%27s_law

9. "Murphy's laws origin - Excerpted from *The Desert Wings*, March 3 1978" Murphy's laws site Accessed September 5, 2015 http://www.murphys-laws.com/murphy/murphy-true.html

Modernizing the Safety Process: Complexity Theory for Flight Test Professionals

Daniel Javorsek II
Air Force Test Center
Edwards AFB, CA 93524

Like all fields of research, flight test continues to evolve as new information and techniques are introduced. Many times this change happens gradually, leading to subtle overall improvements such as those associated with data collection and analysis. However, with respect to safety, the changes are often characterized by discrete responses to specific catastrophic events. After a brief historical summary of worldwide and USAF fatal aircraft accidents we demonstrate that in the modern era, human agency has been a dominant causal factor. This has lead to an effective plateau in the mishap rate that seems unlikely to change unless we shift to better prepare for inevitable but unpredictable events. As has already occurred in other fields, we recommend embracing the more probabilistic nature of events where uncertainty reigns supreme. We may then develop lessons from Complexity Theory, a field designed to tackle systems whose behaviors are inherently difficult, if not impossible, to predict. Forcing ourselves to acknowledge uncertainty helps us design robustness and makes us better postured to respond to unanticipated, high impact outcomes. The inclusion of human agency extends beyond just the operator but affects test safety planning as well. By investigating the psychology associated with cognitive biases and placing an emphasis on decision making and judgment, we may better shape our perspective. We find that an over reliance on an assumption of Gaussian (Bell Curve) distributed variables can result in the flight test community being surprised more often than expected due to an underestimation of the probabilities associated with fatal mishaps. As a result, we propose changes to the familiar Risk Assessment Matrix to align it better with modern risk management methods and highlight how application of the Partitioned Multiobjective Risk Method might assist us in making more informed flight test execution decisions. Finally, we end with a brief discussion of critical slowing down and introduce that it serves as an example of how methods developed out of Complexity Theory might be adapted to flight test. Shifting to a more probabilistic mindset will help us build in robustness to prepare for human agency. Continuing this educational outreach and line of research so that it modifies the way we approach risk assessment and management, is our best hope of modernizing the flight test safety process.

Introduction

In 1905, Albert Einstein published his four *Annus Mirabilis* (from the Latin for "extraordinary year") physics papers that challenged many pre-existing understandings and ushered in a new era of scientific thought. These papers radically transformed science, its methods, and its worldview, moving it beyond its hitherto conventional deterministic interpretations, and toward an embracing of uncertainty and its role in everyday events. In addition to his revolutionary contributions on special relativity[1] and the equivalence of mass and energy[2], his paper on the photoelectric effect[3] introduced the concept of light quantization that later contributed to the foundations of quantum mechanics. Coupled with his work on Brownian motion,[4] quantum mechanics would evolve into a theory that upset the previously accepted deterministic view of nature to such a degree that even Einstein himself could not accept the result.[5] In fact, in response to this probabilistic representation of reality, Einstein has often been quoted as saying, "God doesn't play dice with the world."[6] In this interpretation, the triad of determinism, reductionism, and stationarity have evolved into a view of a future characterized by probabilities, the observation that the sum of parts produces emergent behavior, and the study of rare, high-impact events.

In addition to the probabilistic applications in quantum mechanics, Einstein's paper on Brownian motion (a transport phenomenon caused by random collisions within a fluid) has been applied far beyond physics and is, for example, a foundation of quantitative finance. The adaptive nature of dynamic feedback and non-linear coupled behavior associated with real world systems has also spurred the more recent development of Complexity Theory. This nascent field blends the probabilistic physics concepts that took shape at the turn of the 20th century with the challenges of real-world scenarios, and has seen significant growth and utility in solving problems in physics, economics, medicine, ecology, and meteorology.[7] In the last decade, the advancement of computing capabilities has enabled the inclusion of uncertainty in complex adaptive system models and which may have lessons that can be translated to flight test.

Flight test continues to evolve as new information and techniques are introduced so the introduction of Complexity Theory may be well founded. Many times generic change happens gradually, leading to subtle overall improvements such as those associated with data collection and analysis. However, changes in the flight test safety process are often characterized by discrete responses to specific catastrophic events. Many of these events have resulted in significant improvements to aircraft reliability and maintenance, such that human agency now dominates overall air vehicle system uncertainties. While the current flight test safety risk assessment approach has been very effective at reducing the absolute number of fatal accidents, it is in need of an update. It includes a relatively naïve incorporation of probabilities consistent with those utilized by the 19th century deterministic view that dominated scientific thought at that time. Unfortunately, we are repeatedly reminded that collective human responses and behavior are difficult to predict and model. It should be no surprise that this is not a characteristic unique to flight test. As has been done in other fields, by shifting our focus to a more probabilistic approach, we might be able to better understand and represent this uncertainty.[8]

By interpreting an air vehicle (with the aircrew included) as a complex adaptive system, Complexity Theory provides a mechanism to predict unanticipated emergent macroscopic behavior, highlighting the role of uncertainty in flight test safety, risk analysis, and program management. Flight test events have both deterministic and random elements; an overly deterministic safety mindset underestimates the probabilities of the most catastrophic events and actually increases risk by leading test teams into a false sense of security.

One artifact of the current deterministic mindset is the common tendency for mishap investigation boards to routinely conclude that the catastrophe could have easily been avoided if only someone had "connected the dots" or broken a link in the causation "chain". This fallacy ultimately arises from the retrospective predictability of such events. Unfortunately, we are lured into this conclusion due to our lack of adequate understanding of the time-asymmetry associated with a world heavily influenced by uncertainty and randomness. Armed with this knowledge, we take a fresh look at mishap risk assessment and investigate potential methods to modify it by introducing techniques which allow for accommodations to assess ordinary and catastrophic accidents differently. Adopting this will require a paradigm shift[9] in how flight test safety, analysis, and program management are viewed and implemented.

The Mishap Rate Plateau

In the early days of flight it was understandable that mechanical failures dominated catastrophic mishaps. Additionally, in the absence of modern flight data recorders, post flight analysis, and reconstruction methods, it was often difficult to determine the mishap's cause. However, over time, our understanding of the science of flight improved as we gathered more data on both aircraft systems (aerodynamics, propulsion, structures, handling qualities, etc) and the flight environment (weather, planning, reporting, etc). This led to design and maintenance guidelines which resulted in more reliable aircraft, as well as better flight planning, both of which contributed to a reduction in the fraction of mishaps attributable to them. In contrast, beyond pilot training, little has been done to reduce the uncertainty associated with human agency and as a result, the relative fraction of fatalities from human errors has increased.

This is supported by data obtained from the Bureau of Aircraft Accidents Archives (BAAA) which is a repository of worldwide accidents since 1918.[10] Figure 1a, presents the top three causal factors in fatal aircraft accidents from 1918 to 2014. In an effort to assess the role of human error in US Air Force fatalities, an exhaustive study of 373 Class A mishaps was performed based on accident investigation board results available from the Judge Advocate General (JAG) legal operations agency claims and tort litigation Accident Investigation Board (AIB) repository.[11] Unlike the BAAA archives, the JAG data does not list accident causes, so each AIB report had to be read and studied in order to determine the causal factors, based on the AIB president's concluding remarks. The results provided in Figure 1b show a similar trend, where human agency dominates recent USAF fatal accidents.

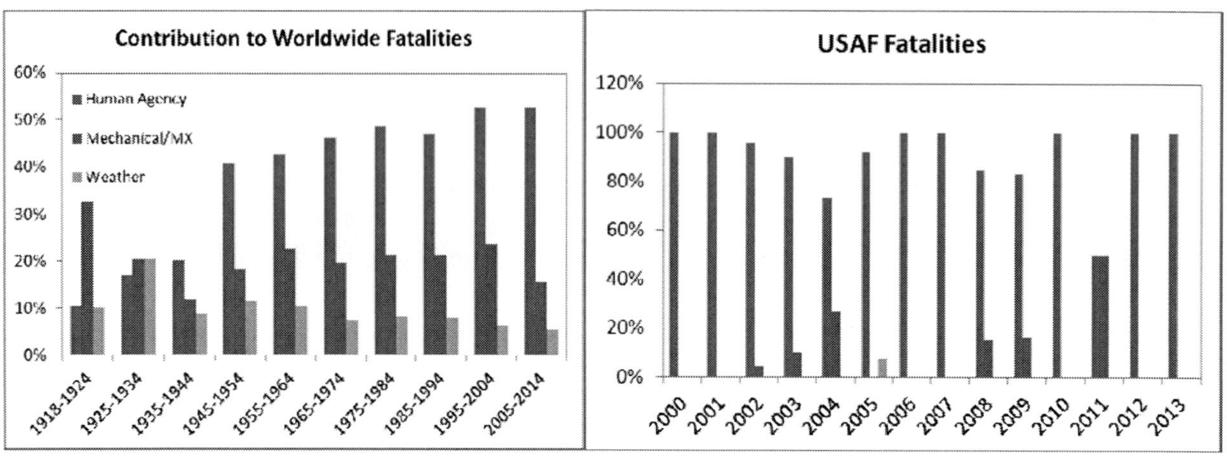

(a) (b)

Figure 1. Top three contributing factors in (a) worldwide fatalities from 1918-2014 and (b) USAF fatalities from 2000-2013.

In the USAF Safety Center's Annual Report for Fiscal Year 2012, Major General Margaret Woodward made the observation that, "The last 10 years are marked by further mishap reduction, but those mishaps were costly – \$8.8 billion and 767 lives. And while the last five years show continued improvement, the trending data illuminate a new concern – we've plateaued."[12] The statement that the data suggest a plateau is based on an essentially level fatality rate (around 0.4 fatalities per 100,000 flight hours) and a level aircraft destruction rate (around 0.5 aircraft destroyed per 100,000 flight hours) over the last decade.

While this may initially seem surprising when viewed in the context of sources of uncertainty, the aviation community has effectively controlled and mitigated most outside factors. Although large strides have been made over the last century with respect to weather and maintenance (often by implementing procedures that inherently reduce risk, as well as timing inspections to catch mechanical failures before they occur), little has been done to change or control pilot actions and responses. Pilot training and education have certainly improved, but human agency remains a source of considerable uncertainty. This uncertainty often stems from subconscious cognitive biases associated with limitations in our neurobiology and is a topic we will expand upon in sections that follow. Although the aviation community has attempted to correct for this with a variety of measures (including regular education programs on Crew Resource Management and Operational Risk Management), the fact that it is a subconscious artifact of normal brain functioning makes it a strong source of uncertainty.

In the end, the uncertainty associated with human agency is not limited to the obvious ones related to the operator but rather extend to the entire test team which includes those involved in program management and risk assessment. This creates a set of conditions that suggest the broader flight test community may be able to learn from recent progress made in Complexity Theory, which is better equipped to account for such uncertainty.

Complexity Theory 101

Complexity Theory emerged to address systems comprised of heterogeneous, interdependent, interconnected, and adaptive agents. The inclusion of adaptive agents is one of the distinguishing qualities of this field and what drives a slight shift in nomenclature from "component" to "agent." Systems common to physics and engineering often possess some of the other qualities (notably the interdependence and interconnectedness) but often are unable to incorporate a system that is adaptable. In fact, aerodynamics and survivability engineers are intimately familiar with coupled, non-linear equations of state as evidenced by their mastery of the Navier-Stokes or Stratton-Chu equations. For example, an aircraft and its resulting aerodynamic/electromagnetic properties is a complicated system that is made from components that obey well defined rules. In principle, the aircraft is a complicated sum of its constituent parts and accurate predictions may be difficult, but are possible. This quality is the main reason for the improvement in the mechanical and maintenance contribution to fatal mishaps (see Figure 1a). A complicated system enables prediction and characterization of the system as a whole by understanding the governing behavior of each component part (no matter how small); an assumed property of natural systems which dominated 19th century scientific thought. In such systems, residuals and perturbative sensitivity studies have shown that Gaussian or Normal distributions (colloquially referred to as the Bell Curve) are adequate.

In contrast, human-introduced adaptability makes prediction difficult because a well-defined set of state equations is impossible. If an aircraft would be a good example of a complicated system, air traffic would be a good example of a complex one. In econometric complex adaptive system models, the human agency is introduced via a set of "rational" actor assumptions that derive their roots from Game Theory (which is sometimes called Interactive Decision Theory to avoid the pejorative connotations typically associated with the name).[13]

While this remains incomplete, it is a first order attempt to include and model human behavior. Unfortunately, many surprises persist when investors fail to remain "rational" in their judgment and decision making. As a result, economists sometimes incorporate an adjustment approximating the uncertainty from human agency which works well in theory since the underlying distribution is known a priori. However, this is not the case in practice, since economists frequently assume a Gaussian distribution function since they are unable to ascertain the truth from a very small event rate. Use of the wrong distribution leaves them vulnerable and gives rise to many of the unexpected and dramatic behaviors (such as the 2007 financial crisis).

In complexity, the interdependent behavior of connected, heterogeneous, and adaptive agents frequently results in emergent macroscopic features not predictable from fundamental laws. These concepts underlying complexity are not new; the "invisible hand" introduced in 1776 via *The Wealth of Nations* by Adam Smith is an early description of such emergence, since well-defined order arose from the social behavior of butchers, brewers, and bakers.[14] However, a more formalized development had to await the advancement of appropriate computing capabilities.[15]

As the speed and ease of computation advanced, Complexity Theory has taken shape in a variety of diverse fields. Since complex systems, and the emergent phenomena which often accompany them, arise from dynamic and competing feedback mechanisms, the emphasis has shifted from the idea of direct control to exercising influence. In a complex system no single agent can control the end state but every agent can, at some level, influence it. In such systems, uncertainty in initial conditions can have dramatic effects on the final end state. In fact, Chaos Theory popularized in the 1990's (and which is a subclass of Complexity Theory) punctuates this behavior. Uncertainty and probability distribution functions play a key role in understanding any complex adaptive system to include an air vehicle with the aircrew and test team included.

Modeling Uncertainty

To help frame the mindset of the 19th century it is helpful to capture the perspective that predated the probabilistic transformation in the natural sciences. This is best expressed in the statement made by Marquis Pierre Simon Laplace in 1825:

> We may regard the present state of the universe as the effect of its past and the cause of its future. An intellect which at any given moment knew all of the forces that animate nature and the mutual positions of the beings that compose it, if this intellect were vast enough to submit the data to analysis, could condense into a single formula the movement of the greatest bodies of the universe and that of the lightest atom; for such an intellect nothing would be uncertain and the future just like the past would be present before its eyes.[16]

One measure of the consensus views regarding the level of determinism that persists in flight test may be found in the typical response to a mishap. By analyzing both the formal safety board results and the common responses of line test pilots when confronted with the failures leading to an incident, we discover a remarkably deterministic view reminiscent of Laplace's 1825 statement. As in many unrelated cases, the cause and effect relationships that look good on paper or in an after-action report often misrepresent the true complexity of real-life events rife with pressures and influences that are difficult, if not impossible, to quantify either before or after an event. Whether this arises for fatal aircraft mishaps or events such as Pearl Harbor or 9/11, this often leads to the proverbial case where there is a perception that individuals involved in the event "failed to connect the dots" that clearly (in retrospect) indicated the potential for disaster.

This rhetoric inherently implies a mindset where events transpire predictably from past to present and are thus controllable. At the other end of the spectrum there are instances where the role of uncertainty in high-impact events occurring with low-frequency are grossly overstated, implying a world that is random and beyond our influence. For example, David Viniar, the Goldman Sachs chief financial officer, stated of the 2007 financial crisis, "We were seeing things that were 25-standard deviation moves, several days in a row."[17] This statement highlights the incorrect reliance on use of the Gaussian distribution to represent a clearly non-Gaussian event. For example, the time required to observe the 25 standard deviation "move" referred to by Mr. Viniar exceeds the current age of the universe (also recall that the standard deviation is a property unique to the Gaussian distribution providing conclusive evidence of the statistics he is utilizing).

Additionally, this overly deterministic view fails to recognize the time asymmetric nature of many flight test mishaps. This asymmetry leads to what is often referred to as "retrospective predictability" or "retrospective distortion" since, when viewed in reverse, the event appears straight forward to predict.[18] To best demonstrate this time asymmetry it is convenient to consider a single particle of pollen floating in a glass of still water. After several hours of random collisions with adjacent water molecules (applying this Brownian motion generically is often referred to as a random walk[19]), the pollen will have traveled about an inch under normal conditions.[20] In Figure 2, the pollen's motion is governed by inelastic collisions with water molecules in the glass such that the total distance traveled, D, is given (in two dimensions) by $D = L\sqrt{n}$, where n is the number of collisions and L is the average distance between them. If observed, it is possible to filter the collective motion of all the water molecules and focus only on those that are relevant. This then enables an accurate explanation of each collision that ultimately led to the pollen's current location (the path from P_1 to P_2 in Figure 2 could be readily explained). In contrast, an attempt to predict the future location of the pollen (P_3 in Figure 2) must account for all possible paths and collisions. Prediction becomes impossible due to an unrealistic number of calculations far greater in scope than those associated with the aforementioned observation assessment.

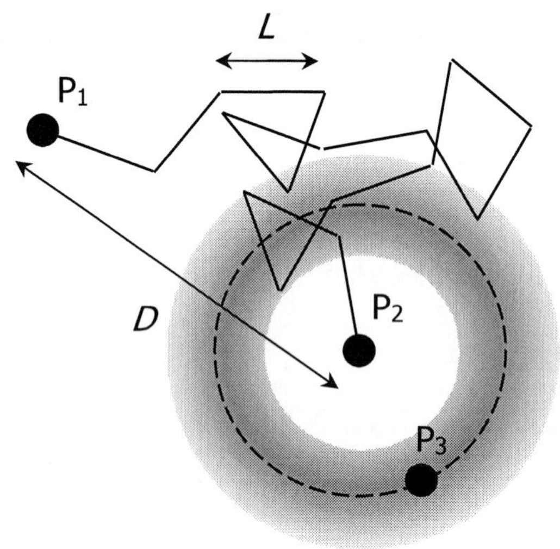

Figure 2. The random walk motion of a pollen particle suspended in still water is a simple example that represents the time asymmetry of events.

Even in this rather straightforward example from physics absent of any adaptive influences, prediction is relegated to generating a probability distribution function based on our knowledge of the system dynamics. In complicated physical systems, experiments can yield high event rates that permit verification of theoretical distribution function validity. As with many simple examples, the distribution function for the future position of the pollen can be represented by a Gaussian since the velocity distribution of the water molecules are represented by the Maxwell-Boltzman distribution which approximates a Gaussian at moderate temperatures.

The deterministic theorist (i.e., Laplace's intellect mentioned above) might conjecture that with enough computing power and sufficiently accurate knowledge of the initial conditions calculation of all possible paths should be possible. However, to demonstrate the intractability of the pollen problem consider the even more simple example of a caroming billiard ball on a frictionless pool table littered with identical balls. The more impacts to be predicted require ever more precise knowledge of the initial conditions (each ball mass, each ball location, gravitational forces on each ball, etc.). While the first impact is fairly straight forward, each successive collision the calculation gets exceedingly complicated, such that by the ninth impact, the gravitational force of the player must be accounted for, and to compute the 56th impact, every particle in the universe needs to be included.[21] If prediction for such a simple example is this challenging, it is safe to say that prediction involving uncertainty at the levels of those associated with flight test are only possible with probability distribution functions.

In introductory statistics courses a large emphasis is placed on the Gaussian distribution because it represents a simple example from which to build upon. The Gaussian distribution's applicability relies upon the Central Limit Theorem, which assumes that variables are independent and random with well behaved expectation values and variance. In contrast, Heavy Tailed distributions are non-Gaussian such that the extremes are not exponentially bounded (see Figure 3).

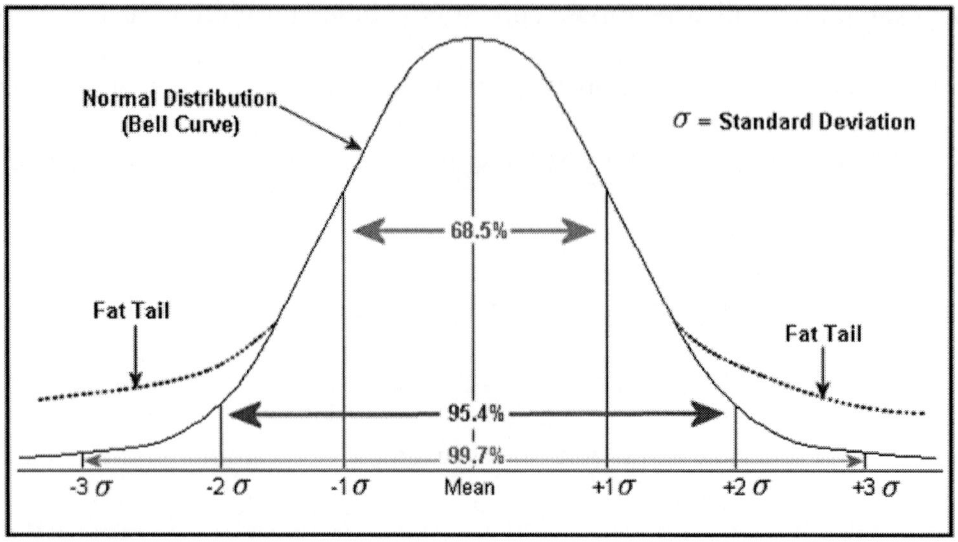

Figure 3. Comparison of the Gaussian or Normal Distribution (Bell Curve) and a class of distributions commonly referred to as Heavy Tailed distributions.

In Heavy Tailed distributions (examples include the power-law, log-normal, Levy Flight, Cauchy-Lorentz, etc.) the variance is not finite as it is for the Gaussian.[22] In economics and finance this is extremely undesirable because it implies considerable extra risk. In flight test, the same is true, such that "improbable" events based on a Gaussian may actually be expected if the variables of concern are actually Heavy Tailed. In the end, ignorance of the underlying distribution can lead us to severely underestimate the probability of extreme events and assuming the variables are Gaussian, when they are actually Heavy Tailed, can lead us to be under prepared for mishaps.

To see the potential impact of a Heavy Tailed distribution we expand on the previous pollen example. Figure 4a provides the hypothetical two-dimensional (2D) motion for pollen dominated by collisions that follow a Gaussian distribution and Figure 4b shows those which follow a subclass of Heavy Tailed distribution called a Levy Flight. While both start at the same origin and run through 1000 subsequent collisions, their resulting paths are dramatically different. For example, compare the scale of the two figures, as well as the more chaotic nature associated with the big jumps that are indicative of the Heavy Tailed Levy Flight model. This example using the Levy Flight distribution has many large jumps to emphasize the impact but this is not a requirement of all Heavy Tailed distributions. In fact, many Heavy Tailed distributions look much more like the Gaussian most of the time but still retain the potential for high impact, rare events. In our analogy with flight test, the long straight lines in Figure 4b would represent unexpected surprise events due to human agency that would not be predicted assuming the well behaved case in Figure 4a. Also recall that only a single observation is needed to reject the Gaussian because it does not allow for large deviations. As a corollary, billions of observations will not fully confirm the validity of its application because Heavy Tailed distributions do not prohibit long quiet Gaussian-like stretches.

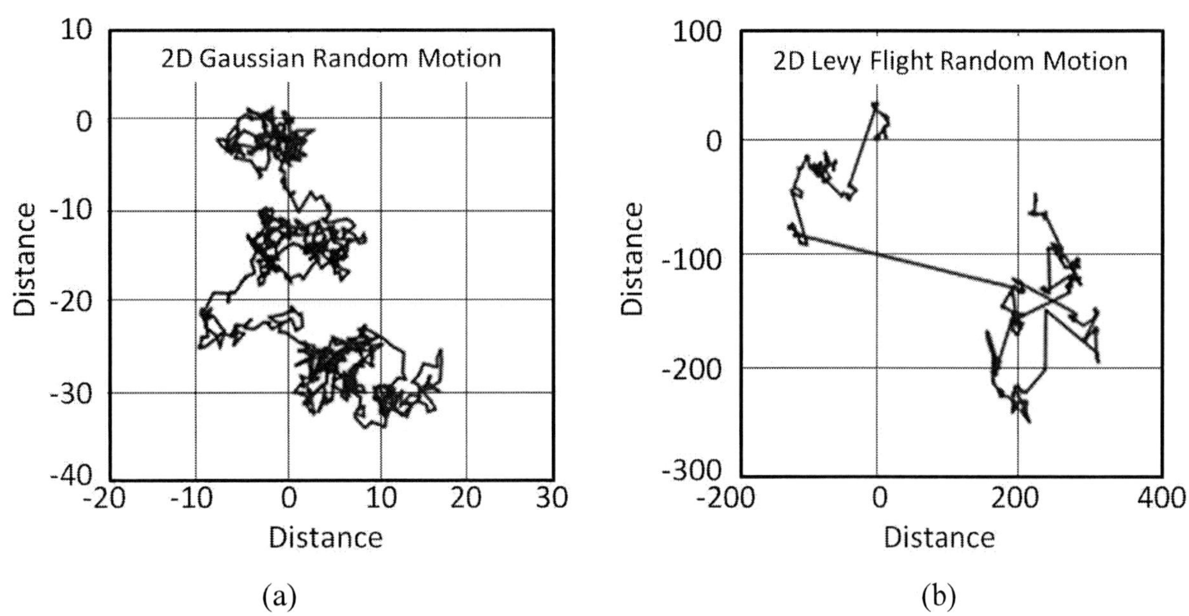

Figure 4. The 2D motion of a pollen particle suspended in water with collisions that follow both the (a) Gaussian distribution and the (b) Levy distribution.

The goal of this discussion is not to encourage us to suddenly change many of our statistical representations to be Heavy Tailed ones but rather to introduce skepticism. The notion of the Heavy Tailed distribution helps to mitigate risk by making high impact, unexpected events conceivable. We should not strive to design for every possible outcome but instead confine the negative aspects of human agency so that an execution error doesn't result in a fatal outcome.

Additionally, while these discussions about distribution functions are useful in an academic context they do not translate well to the real world. This is mainly because in the real world we observe events and not probability distributions. Unlike the frequently casino games which have well known probabilities and rules in flight test we do not know the odds but instead must discover them when the sources of uncertainty are not defined. As a result, mishap boards often dissect events that happened and forget to consider those events that could have happened but did not. This artifact of the determinism mindset can result in the loss of clues that may be extremely vital to our assessment of overall system robustness and fragility. Robustness sometimes arises from system redundancy but is not without significant cost.

In program management, redundancy cost-benefit decisions are routinely made based on the assumption of risk that follows from a risk assessment method heavily dependent on the anticipated (or estimated) probabilities. If a given undesirable outcome is sufficiently remote, a fiscally constrained program manager might elect to permit or encourage an "optimized" design that is less robust to failure. As a result, the assumption of a Gaussian which predicts exponentially unlikely deviations would lead to an underestimation of probabilities and could unknowingly create an aircraft less resilient to failure.

Fatal catastrophic mishaps are, by definition, not the result of well-behaved variables.

Thus the Central Limit Theorem and the Gaussian distribution inadequately describe them; they instead arise from Heavy Tailed distributions and are impossible to accurately predict. Which Heavy Tailed distribution they belong to is unimportant and due to the low event rate impossible to determine. What *is* important is that we understand we are unable to accurately estimate their probability and thus do not permit a reduction in robustness and resiliency. Armed with this knowledge we can make some recommendations on how to modify our risk assessment methodology.

A Fresh Look at Risk Assessment

Current safety risk assessment methods are an excellent first attempt to incorporate a probabilistic mindset but need to be modified to account for the uncertainty in probabilities detailed in the preceding section. While publications such as the FAA Risk Management Handbook[23] and the USAF AFTC Test Safety Review Policy[24] provide good qualitative guidance, there are no discussions with regard to the potential to underestimate mishap probabilities. These instructions instead rely heavily on "engineering judgment." This is possibly another cause of the plateau in the fatal mishap rate observed over the last decade.

The reliance on "engineering judgment" suggests the need to investigate the psychology of estimation, for which there is an abundance of research. Beginning in the 1970's Amos Tversky and Daniel Kahneman began significant research to evaluate the psychological effects and cognitive biases associated with judgment, self-assessment, estimation, and prediction as they apply to situations that contain a high degree of uncertainty.[25] They routinely confirmed that humans regularly overestimate their confidence and are subject to a variety of biases for which a reduced list is provided in Table 1.

Table 1. Summary list of Cognitive Biases important for flight test professionals.[26]

Cognitive Bias	**Description**
Ambiguity Effect	The tendency to avoid options for which missing information makes the probability seem "unknown."
Anchoring/Focalism	The tendency to rely too heavily, or "anchor", on one trait or piece of information when making decisions (usually the first piece of information that is acquired).
Automation Bias	The tendency to excessively depend on automated systems which can lead to erroneous automated information overriding correct decisions.
Bandwagon Effect/ Groupthink/ Herd Behavior	The tendency to do (or believe) things because many other people do (or believe) the same.
Base Rate Neglect/ Conjunction Fallacy	The tendency to ignore base rate information (generic or general information) and focus on specific information (information only pertaining to a certain case). Assume that specific conditions are more probable than general ones.
Confirmation Bias/ Expectation Bias	The tendency to search for, interpret, focus on and remember information in a way that confirms one's own preconceptions. Additionally to disbelieve, discard, or downgrade the corresponding weightings for data that appear in conflict with those expectations.
Halo Effect	Observer's overall impression of a person, company, brand or product influences observations downplaying data to the contrary.
Overconfidence Effect	Excessive confidence in one's own answers to questions. For example, for certain types of questions, answers that people rate as "99%" certain turn out to be wrong 40% of the time.
Regressive Bias	A certain state of mind wherein high values and high likelihoods are overestimated while low values and low likelihoods are underestimated.
Retrospective Distortion	The tendency to see past events as being predictable at the time those events happened.

Without going into great detail on each of the biases in the table it is useful to look into one simple example, such as "anchoring", whereby an estimate is highly dependent on the initial value, its anchor. In such experiments, subjects were first shown a random number (by spinning a wheel of fortune which they knew to be random) and were subsequently asked to estimate something for which they did not know the correct answer. Multiple related experiments have confirmed that the random number repeatedly and subconsciously influenced estimates, even when controlled for by any number of demographics to include education and occupation.[27]

The overconfidence effect is similarly well documented and results when one's subjective confidence in a judgment is greater than one's objective accuracy.[28] Overconfidence manifests itself as overestimation, overplacement (excessive performance with respect to others), and/or overprecision (excessive certainty). For example, think of how many fellow pilots would rate themselves in the bottom half of any probability distribution of all pilots at landings (or any task for that matter), yet some portion of the overall population must make up that part of the distribution. Ultimately it is simply human nature (hubris) that creates a large disparity between what people think they know and what they actually know. This should make us skeptical of relying almost exclusively on "engineering judgment" when estimating the probability of fatal mishaps.

This uncertainty in our ability to properly estimate a fatal mishap's probability combines with the non-linear nature of the columns in the standard Risk Assessment Matrix (see Figure 5). In addition, the Risk Assessment Matrix treats all causes equally and does not explicitly handle human agency differently even though we have shown that our ability to accurately represent the probabilities are strikingly more difficult than say evaluating the mean time between failure of a mechanical component. Starting from the right hand side, a change from a Negligible to a Marginal rating and from a Marginal to Critical rating are often dominated by an incident's monetary cost or by the qualitative assessment of the injury (from superficial to minor to severe).

RISK ASSESSMENT MATRIX				
SEVERITY / PROBABILITY	Catastrophic (1)	Critical (2)	Marginal (3)	Negligible (4)
Frequent (A)	High	High	Serious	Medium
Probable (B)	High	High	Serious	Medium
Occasional (C)	High	Serious	Medium	Low
Remote (D)	Serious	Medium	Medium	Low
Improbable (E)	Medium	Medium	Medium	Low
Eliminated (F)	Eliminated			

Figure 5. Standard flight test safety risk assessment matrix.

From the aircrew's perspective there is likely little difference between a "superficial" and a "minor" injury. While a larger difference exists between a minor injury and a severe one, the step from there to death is quite large (from the aircrew's point of view it is infinitely severe). This non-linear severity change is graphically represented in Figure 6 for both the cost scale and the aircrew injury scale. These different internal relationships between cost and aircrew injury represents a multiobjective problem whereby we are unable to treat them equally or with the same process. For example, the cost can use the traditional Bayesian expected value for the risk defined as the product of the probability and the severity. In contrast, the fact that a fatality is a potential outcome for the catastrophic aircrew injury case means that the Bayesian expected value calculation breaks down. More on this point later.

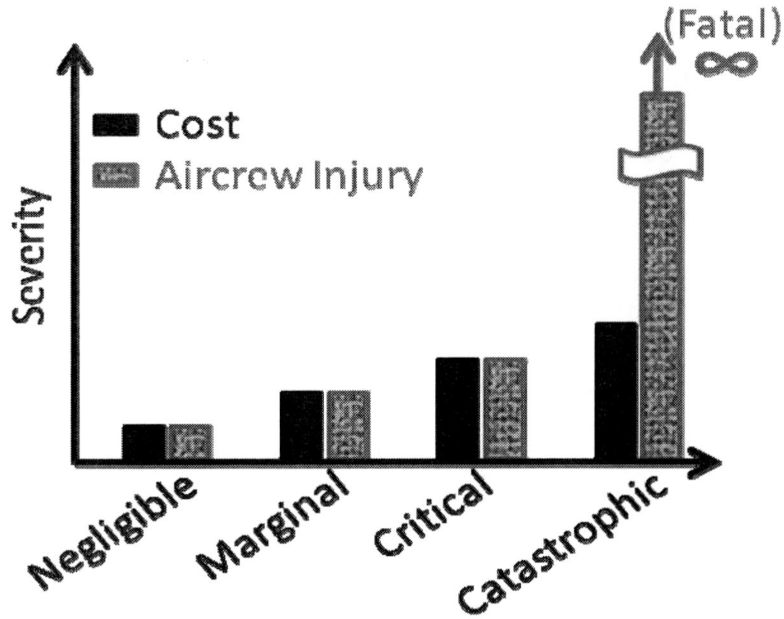

Figure 6. Relative scale for the different Risk Assessment Matrix severity ratings.

In the end, the Risk Assessment Matrix and the resulting overall risk assessment are very dependent upon both the probability estimation and the illusion of equal weighting. In fact, from both the program management and risk acceptance perspectives, decisions are regularly made on how much testing is enough or whether a given test point is too risky, based upon this assessment. Unfortunately, fatal accidents have historically shown us that we may not have done the cost-benefit analysis properly as we second guess our decision with retrospective predictability.

One of the fallacies that continuously affect our decisions is how we interpret past data. In a world governed by Gaussian distributions, confidence in extrapolation into the future is bolstered by successes due to each outlier's low probability and ultimately a matter of incomplete information. For example, consider the life of a Thanksgiving turkey. Initially, he is concerned about the farmer and his intentions. However, after three years of being fed every morning, safety is apparently confirmed, and on Thanksgiving morning he boldly, yet naively predicts the same positive outcome. If the turkey instead lived in a world governed by non-Gaussian distributions he might have a more shrewd outlook and view each feeding as confirmation of the persistent danger of becoming dinner.

Since highly unexpected events have occurred in flight test when past data suggested that a maneuver or technique worked means that the shrewd outlook of the Turkey may be more appropriate. In short, high risk test points accomplished uneventfully in the past provide no guarantee of future risk reduction and, to the contrary, may mean that we are simply primed for the conditions to change just enough that a mishap is imminent. The intent of this discussion is not to make us paranoid but rather to point out our own limitations and biases as well as to honestly acknowledge and identify uncertainties. To use the words of Nassim Taleb, "It is much more sound to take risks you can measure than to measure the risks you are taking."[29]

Given our cognitive biases with regard to overestimation and our inherent inability to correctly assess the probability of extreme events, we suggest a change may be in order. It might be prudent to consider adjusting the Catastrophic column of the Risk Assessment Matrix such that it is always considered High risk regardless of the estimated probability. This idea that risk management should be partitioned was pioneered in the mid-1980's by Eric Asbeck and Yacov Haimes in response to the fact that the Bayesian method of expected value provides the same end result for low probability, high cost event as a high probability, low cost event.[30]

As a result, Asbeck and Haimes introduced the Partitioned Multi-objective Risk Method (PMRM) which differentiates between the two by creating a conditional expectation of events and specifying partitioned probabilities for different objectives. As a result, conditional expected-value risk functions are generated for each of the predefined ranges, which are then augmented in the original optimization problem as new objective functions.[31] Before PMRM was developed, problems with at least one random variable were solved by computing and minimizing the unconditional expectation of the random variable representing the event. In order to show how this might be applicable to flight test we will postulate a simplistic first flight example.

A test team is trying to decide between accomplishing their first flight event at a remote desert location or at the urban contractor facility. The tradeoffs in this simplistic example are between the additional cost associated with transporting and operating the vehicle from the remote location versus risking a catastrophic loss over the more densely populated city surrounding the contractor facility.

In typical risk assessment the expected loss of an accident, L, is represented by the sum of the product of the probabilities of occurrence, P_A, with the associated costs of each accident, C_A, such that:

$$L = \sum_{A} P_A C_A \,.$$

Using completely notional numbers, let's assume that $P_A = 1/10{,}000$ and the $C_A = \$75M$ for a crash in a remote location (i.e. the cost is primarily that of the aircraft) but $C_A = \$500M$ for a crash over a populated area (i.e. the cost of the aircraft plus buildings, insurance payments, lawsuits, collateral damage, etc). We also might assume that to minimize the impact of such events the test team will take advantage of the fact that the contractor facility is on the edge of the city, such that the first flight will only spend 1% of the flight time over populated areas. As a result, we might compare the associated expected loss for each scenario to arrive at the following:

$$E_{remote} = \left(\frac{1}{10{,}000}\right)(\$75{,}000{,}000) = \$7{,}500$$

$$E_{contractor} = \left(\frac{99}{100}\right)\left(\frac{1}{10{,}000}\right)(\$75{,}000{,}000) + \left(\frac{1}{100}\right)\left(\frac{1}{10{,}000}\right)(\$500{,}000{,}000) = \$7{,}425 + \$500 = \$7{,}925$$

Given this methodology, a fiscally savvy program manager might be encouraged to decide that the $425 difference isn't worth the cost of making the change to move flight testing to the remote location. However, using PMRM one would argue that even though crashing into a populated area is unlikely, it is also totally unacceptable. PMRM argues that decisions should be based on both the expected normal loss and the expected catastrophic loss, which results in two objectives (thereby introducing the multi-objective aspect into the decision making process).

Although there are many ways to apply PMRM to a variety of situations we suggest that this reinforces our prior proposal to adjust the familiar Safety Risk Assessment Matrix by holding that any outcome with a catastrophic severity be treated differently, such that no matter what the probability the risk is always High (instead of allowing it to graduate to Serious or Medium as probability decreases).

Coupling this with our discussion in the preceding sections regarding the difficulty associated with accurately predicting the probability of catastrophic events might bring a welcome change to help break the current mishap rate plateau. However, as demonstrated in the first flight example above, this will not be as convenient or cost effective as our current approach. It will inherently cause more oversight since more missions would be subject to the additional scrutiny afforded higher risk events. If this forces the test team to more aggressively pursue alternative test points or increases the time spent simulating and critiquing past maneuvers associated with the event this may be an important development in flight test safety's evolution.

In addition to the recommended changes to risk assessment and management, which largely arise from the necessity to include uncertainty when discussing complex adaptive systems, Complexity Theory has much more to offer the flight test community. One specific non-human agency example which may apply to test safety (via handling qualities, flying qualities, flutter, etc.) is derived from recent work generating universal early warning signs for critical transitions.

Early Warning Signs for Critical Transitions: An Application from Complexity Theory

Phase transitions have been studied for centuries and applications of the general thermodynamic formalism goes beyond that which is taught in introductory science classes. A wide range of systems display behavior governed by highly coupled and non-linear differential equations of state. To name a few, these include the familiar mathematical expressions for fluid dynamics (as in the Navier-Stokes equations), electromagnetics (as in Maxwell's equations and the many variations therein), quantum field theory (as in the Yang-Mills, quantum electrodynamics, and quantum chromodynamics equations), and relativity (as in Einstein's field equations). While these state equations can be very successful at dealing with problems that are well defined within the scope of each equation set, many dynamics encountered in a non-laboratory environment experience additional difficulties which the equations do not anticipate. Often such real world systems are very dynamic, with interactions, stochastic influences, and system responses that do not naturally follow from basic principles. Such complex systems have received increased attention since improving computational capabilities have made the field of Complexity Theory more tractable.

In the end, many important complexity problems revolve around predicting the state of a system made of many interconnected elements exhibiting a wide variety of complicated feedback relationships. As the internal organization and relationships change, phase transitions can mark particularly important system-level shifts. Identifying such critical thresholds, colloquially called tipping points, could be extremely useful for a variety of fields. The thresholds for such transitions correspond to bifurcations whereby the system stability rapidly undergoes a sudden change in system response or behavior. As an example from medicine, asthma attacks and epileptic seizures provide very little warning of their impending shift in system response and thus represent a bifurcation. A diverse set of such cases was analyzed by Scheffer et al. which led them to propose that "…the dynamics of systems near a critical point have generic properties, regardless of the differences in the details of each system."[32] These generic properties are:

1. A slower recovery from perturbations
2. Increased variance (conditional heteroscedasticity)
3. Increased autocorrelation
4. Fluctuation asymmetries (skewness, kurtosis, and flickering)

All of these are mathematical properties associated with the system residuals (errors) and are thus applicable to any system. This appears to be generic for a variety of reasons, but at a fundamental level critical slowing down is associated with power-law scaling of the relaxation time (recovery rate). A fast recovery rate implies greater stability and resilience since it is associated with a deeper potential well and provides evidence that the residuals and the system will likely continue to remain well behaved. In contrast, if this recovery rate begins to slow the system may be approaching a transition (see Figure 7) and thus may warn of impending rapid, drastic change. This has been measured for 2D Ising ferromagnetic systems along with medical conditions with bifurcations (epilepsy, asthma, etc.) such that it may be generalized and, as a result, should be applicable to any system with similar properties.

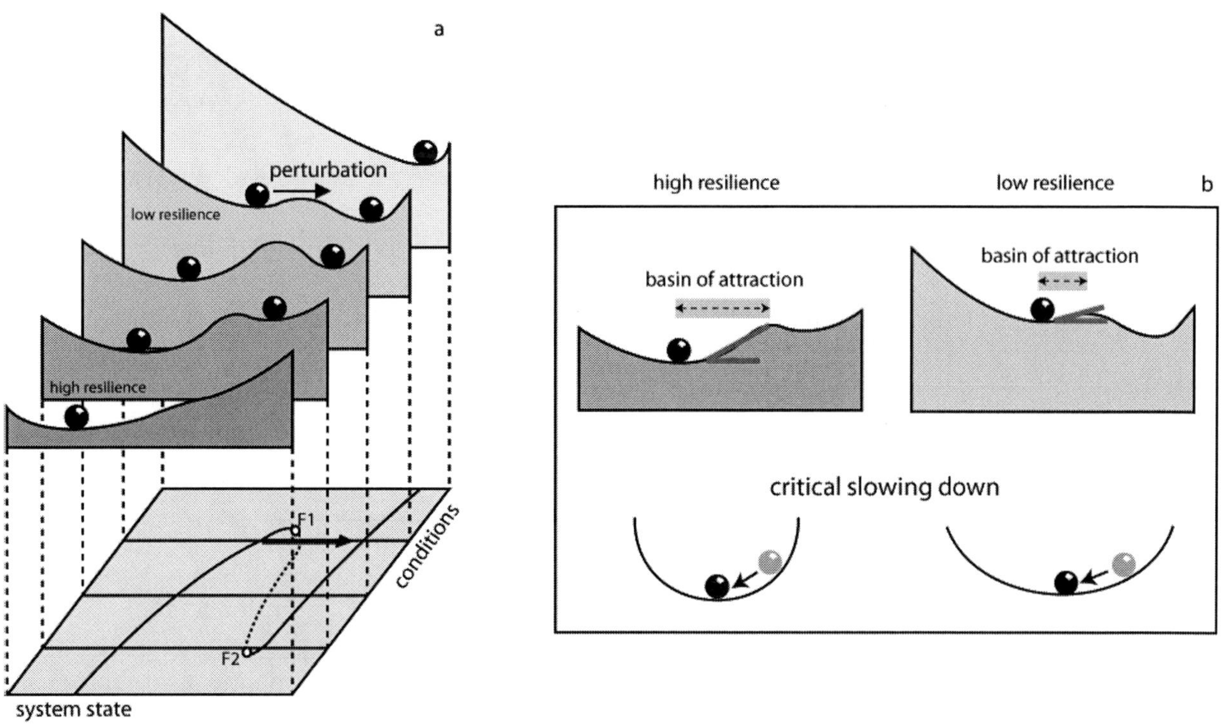

Figure 7. Critical slowing down as a measure of resilience and its relationship to early warning for critical transitions.

In short, these properties have the potential to be a direct indicator of system resiliency and might be useful in a variety of flight test situations. Consider a test program that has accepted some level of increased risk by forgoing the installation of a Flutter Excitation System (FES). Although, considerable modeling and ground vibration testing may be performed, until airborne the model's prediction accuracy (and any inherent risk) remains unknown (as is frequently proven by flight test, all models are wrong…some are useful…some may be dangerous). In this particular situation, there is a need to monitor for potential errors by watching accelerometers, but concern would still persist that sufficient warning would be available to avert disaster. Monitoring for the conditions associated with critical slowing down in real time might provide more advance warning and thus be more robust to human agency, since flutter represents a rather sudden shift in system dynamics.

Similar arguments might be made for an aircraft with suspect flying or handling qualities where divergent flight control regimes may be suspected. While this example does not represent an application which accounts for human agency it does suggest that more may be discovered and further study on how Complexity Theory might be applied to flight test is certainly warranted.

Conclusions

After a brief historical summary of worldwide and USAF fatal aircraft accidents we demonstrated that, in the modern era, human agency has become the dominant causal factor. The resultant effective plateau in the mishap rate seems unlikely to change unless we modify our flight test safety approach. As has already occurred in other fields, we recommend a cultural shift to a more skeptical mindset especially with regard to probability estimation in systems with human agency. We have shown that underestimation of the probability for extreme, yet rare, events can lead to inaccurate programmatic cost-benefit analysis decisions, resulting in less robust systems.

One method to influence this cultural shift would be to improve educational outreach by slightly modifying the curricula of the different Test Pilot Schools (Air Force, Empire, Epner, National, Navy, etc) to cultivate flight test professionals with the skepticism that often only comes with the painful experience of lost friends and colleagues. Expanding beyond introductory level statistics through the incorporation of non-Gaussian distribution functions would help to instill a culture skeptical of precise probability estimates for real-world events influenced by human agency. Additionally, strong consideration should be given to including formal instruction on the psychology of judgment and decision making beyond the abbreviated form currently included in Crew Resource Management, since it would also serve to bolster the desired cultural shift. While it may seem odd to include psychology in a flight test curriculum, consider that one of the best defenses against cognitive biases is awareness that they exist and that they can affect anyone.

In this new culture, we may then develop lessons from Complexity Theory, which is designed to tackle adaptive systems, such as those with human agency whose behaviors are impossible to predict. With complex adaptive systems, the emphasis moves from system *control* to system *influence*. In this context, the more flight test professionals who possess a healthy amount of skepticism, the more they collectively affect the outcome. Each individual nudge, no matter how subtle (such as a simple questioning comment during a simulator session or chance encounter), serves as a correcting perturbation. In a scenario where only one member of a team takes this proactive approach, there is considerable potential that the threshold is not met and the path is not sufficiently altered. However, with more individuals there are more agents able to recognize and correct execution errors, potentially altering the final catastrophic outcome.

Forcing ourselves to acknowledge uncertainty helps better posture us to respond to unanticipated, high impact outcomes. We highlight how errors in probability estimates (either from use of the wrong distribution function or from cognitive biases) can occur in planning and designs that are not as robust as they should be. As s result, we propose changes to the standard Risk Assessment Matrix to better align it with modern risk management methods and highlight how application of the Partitioned Multiobjective Risk Method might assist us in making better informed flight test execution decisions. Finally, we introduce critical slowing down as an example of how such methods, developed out of Complexity Theory, might be adapted to flight test.

In the end, shifting to a more probabilistic mindset skeptical of inevitable human agency will modify the way we approach risk assessment and management. Modernizing the flight test safety process through education and including these changes represent our best hope of breaking the plateau in fatal mishap rates.

Acknowledgements

Thank you to Steve Rainey for his lengthy discussions regarding catastrophic events and the results associated with certain formal mishap boards. These initial conversations helped guide me on the journey that would ultimately lead to my discovery of the field of Complexity. To DIA economist John Schwitz, whose discussions and joint work in Intelligence applications introduced me to Game Theory and agent based modeling. Finally, the quality of this paper would not have been possible without the feedback provided by Brad Carlson and Mike Nielsen who afforded me perspectives that I had not previously considered.

---

[1]Einstein, Albert. "On the Electrodynamics of Moving Bodies." *Annalen der Physik* 17 (1905): 891-921.

[2] Einstein, Albert. "Does the Inertia of a Body Depend Upon Its Energy Content?" *Annalen der Physik* 18 (1905): 639-641.

[3]Einstein, Albert. "On a Heuristic Point of View about the Creation and Conversion of Light." *Annalen der Physik* 17 (1905): 132-148.

[4]Einstein, Albert. "Investigations on the Theory of Brownian Movement." *Annalen der Physik* 17 (1905): 549-560.

[5] Ballentine, L.E. "Einstein's Interpretation of Quantum Mechanics." *American Journal of Physics* 40 (1972): 1763-1771.

[6]Hermanns, William, *Einstein and the Poet: In Search of the Cosmic Man.* Brookline Village, MA: Branden Press, Inc., 1983. 58.

[7]Johnson, Neil. *Simply Complexity: A Clear Guide to Complexity Theory.* Oxford: Oneworld Publications, 2007.

[8]Orrell, David. *The Future of Everything: The Science of Prediction.* New York, NY: Basic Books, 2007. 348.

[9] Kuhn, Thomas. *The Structure of Scientific Revolutions*. Chicago, IL: University of Chicago Press, 1962.

[10] "B3A Aircraft Accidents Archives." *B3A Aircraft Accidents Archives RSS*. http://www.baaa-acro.com/. Accessed May 18, 2015.

[11] "AIB Reports." USAF Judge Advocate General Legal Operations Agency Claims and Tort Litigation, http://usaf.aib.law.af.mil/. Accessed May 18, 2015.

[12] Woodward, Margaret. "Air Force Safety Center Annual Report for Fiscal Year 2012." *AFD 130523-033* (2012).

[13]Osborne, Martin. *An Introduction to Game Theory.* New York, NY: Oxford University Press, 2004.

[14]Smith, Adam. *An Inquiry into the Nature and Causes of the Wealth of Nations.* London, 1776. edited by Andrew Skinner. New York, NY: Pelican Books, 1970.

[15] Miller, John and Scott Page. *Complex Adaptive Systems.* Princeton, NJ: Princeton University Press, 2007. 26.

[16]Davies, Paul and Niels Gregersen. *Information and the Nature of Reality: From Physics to Metaphysics.* Cambridge: Cambridge University Press, 2010. 74.

[17]Larsen, Peter. "Goldman Pays the Price for Being Big." *Financial Times.* August 13, 2007.

[18] Taleb, Nassim. *The Black Swan.* New York, NY: Random House Trade Paperback Edition, 2010. 8.

[19]Mlodinow, Leonard. *The Drunkard's Walk.* New York, NY: Vintage Books, 2008. 197.

[20]Spitzer, Frank. *Graduate Texts in Mathematics: Principles of Random Walk*, 2nd ed. New York, NY: Springer-Verlag, 1976.

[21] Berry, Michael. "Regular and Irregular Motion, in Topics in Nonlinear Mechanics." *American Institute of Physics Conference Proceedings* 46 (1978): 16-120

[22] Grinstead, Charles, and William Peterson, Laurie Snell. *Probability Tales.* New York, NY: American Mathematical Society, 2011. 105.

[23] Federal Aviation Administration. *Risk Management Handbook.* US Department of Transportation FAA-H-8083-2 (2009).

[24] Air Force Test Center. *AFTC Test Safety Review Policy.* Air Force Test Center Instruction (AFTCI) 91-203 (April 30, 2014).

[25] Tversky, Amos and Daniel Kahneman. "Judgment under Uncertainty: Heuristics and Biases." *Science* **185** (1974): 1124-1131.

[26] Pohl, Rudinger. *Cognitive Illusions: A Handbook on Fallacies and Biases in Thinking, Judgment and Memory.* Hove, UK: Psychology Press, 2004; See also Colman, Andrew. *Oxford Dictionary of Psychology*. New York, NY: Oxford University Press, 2003.

[27] Tversky, Amos and Daniel Kahneman. "On the Reality of Cognitive Illusions." *Psychological Review* **103** (1996): 582-591.

[28] Kahneman, Daniel, and Amos Tversky. "On the Psychology of Prediction." *Psychological Review* **80** (1973): 237-251.

[29] Taleb, Nassim. *The Black Swan.* New York, NY: Random House Trade Paperback Edition, 2010. 361.

[30] Asbeck, Eric and Yacov Haimes. "The Partitioned Multiobjective Risk Method." *Large Scale Systems* **6** (1984): 13-38.

[31] Haimes, Yacov. *Risk Modeling, Assessment, and Management.* Hoboken, NJ: John Wiley & Sons Inc., 2009. 482-546.

[32] Scheffer, Marten, et al. "Early-warning Signals for Critical Transitions." *Nature* **461** (2009): 53-59.

Ouch! Let's just make this one a full stop.
Steep Approach Testing on the Citation Latitude

Stuart Rogerson (AF)
Textron Aviation

Scott Foster (AM)
Textron Aviation

ABSTRACT

While conducting abused approach angle testing for steep approach certification for the CE-680A Latitude, a hard landing occurred. This paper will detail the buildup prior to the event, the post incident risk mitigation and the lessons learned while successfully completing the test program.

THE AIRCRAFT

Figure 1 - Latitude CE-680A

The Latitude CE-680A, shown in Figure 1, is a mid size twin engine business jet that can carry up to 9 passengers. It has a 440 KTAS maximum cruise speed with a 2850nm maximum range. The Maximum Takeoff Weight for the Latitude is 30,800 lbs and it was certified in 2015. The Latitude CE-680A aircraft is part of the 680 family. The legacy Sovereign CE-680 was initially certified in 2004 and was certified for steep approach operations in 2006. In 2013, Cessna certified the Sovereign+ which introduced the Garmin G5000 avionics to the Cessna product line. More importantly, the upgrade also changed the outer mold line of the aircraft with the introduction of winglet extensions and canted engine nozzles. Numerous other internal changes were made, but those are not pertinent to this paper. The Latitude is essentially a Sovereign+ with a larger but slightly shorter cabin. The wing, tail, engines and landing gear are identical and maintain the same geometry. The cabin is 6' high with a flat floor. Weights remain similar to the Sovereign+. Figure 2 shows a comparison between a legacy Sovereign and the Latitude. Both the Sovereign+ and the Latitude are pursuing steep approach certification.

Figure 2 - Legacy Sovereign vs. Latitude

Of particular importance to this paper is that this family of aircraft has very similar Sub Part B characteristics. Stall speeds and characteristics, stability and control and performance are all very comparable between the three aircraft. This is true for both an uncontaminated wing and an aircraft simulating icing conditions with ice shapes installed.

STEEP APPROACH

Certain locations in the world require a steeper than normal approach in order to operate in and out of their airport. Per the regulations, a steep approach is defined as any approach which has a glide path greater than or equal to 4.5 deg. For example, London City has a 5.5 deg glide slope due to the geographic constraints of buildings surrounding the airport, as shown in Figure 3.

Figure 3 - London City Airport

As can be seen in Figure 3, this airport is located near the downtown of London. This makes it a highly desirable location for our customers to use since they can easily access the city center. Lugano, Switzerland (Figure 4) is another example of an airport that has a steep approach. In this case, the geographic constraints are the surrounding mountains. Lugano is only accessible during visual meteorological conditions unless you can fly this steep approach. The steep approach is a 6.65 deg electronic glidepath with a transition to a 6.0 deg visual segment for landing.

Figure 4 - Lugano, Switzerland

Steep approach by itself is a challenge; however most airports that are served by a steep approach are also short. In the case of London City, the airport is constrained by the river Thames and is only 4327' long. Lugano is also short at 4658' due to the surrounding mountains and lake. When discussing steep approach it is a two part challenge; being able to fly the steep approach and being able to land the aircraft in as short as distance as possible from that approach.

Steep approach requires not only aircraft approval but operational approval. This operational approval usually is not worth the extra effort for private operators, so the majority of our steep approach customers are commercial operators. Commercial operators are required to use the JAR Ops Landing factors that are 1.67 for a dry runway and 1.92 for a wet runway. If one considers a typical landing distance of 2,500' this becomes 4,800' on a wet runway. However, Textron Aviation is committed to providing reasonable all weather access to London City for our commercial customers, and as we well know London is frequently wet! In order to maximize our performance, Textron

Aviation has previously lowered the flare height from 50' to 35', which requires the use of visual guidance to touchdown. This 15' lower altitude has a significant effect. It reduces the air distance by 250' which when combined with a 1.92 factor makes a difference of 480'; a large reduction on a 4,300' runway.

Textron Aviation's goal for this program was to build on the steep approach certification of the legacy Sovereign, maintaining similar landing performance. This was planned as a post Type Certificate (TC) effort, but for convenience sake some of the data was collected during the pre TC testing. In addition to the legacy Sovereign capabilities at London City, the plan was to add Flight Into Known Icing (FIKI) for steep approaches and to consider steeper approaches such as Lugano.

Steeper angles require higher drag on the aircraft in order to maintain speed. To achieve this, the legacy Sovereign used the normal full flap (35 deg) along with speed brakes. This same configuration was planned for the Latitude. The 680 family has 10 speed brake / spoiler panels, five on each wing. The inboard three panels on each wing are variable speed brakes / spoilers and the remaining four outboard panels are speed brakes which can be either full open or closed. For steep approach, the three inboard panels are fully extended as shown in figure 5. This creates a unique steep approach

Figure 5 - Steep Approach Speed Brake Configuration

configuration for our aircraft. As a result, the approach speed for steep approach is 1.23 times the stall speed for this configuration. This is approximately 6 knots higher than a normal approach speed. Due to the unique configuration, a significant amount of Sub Part B testing is required. This will be described later in this paper.

REGULATORY ENVIRONMENT

Since Textron Aviation is based in the USA, initial certification is done through the FAA. However, when foreign certification is sought, the program is affected by EASA regulations. In the case of steep approach, the majority of the approaches are located in Europe under EASA's jurisdiction and our commercial operators are usually registered with EASA. An important item to note is that the EASA regulations are codified whereas the FAA has Advisory Circular (AC) guidance. Current guidance for steep approach with the FAA is found in AC25-7C– Flight Test Guide for Certification of Transport Category Airplane, paragraph 231. EASA regulations are found in CS-25 Book Appendix Q.

AC25-7C paragraph 231.c(3) states that the landing distance begins at a point 50' above the landing surface. However an Equivalent Level of Safety (ELOS) can be used for a different height to take advantage of precision approach guidance at an airport that provides it, such as London City. EASA is slightly different in that the default flare height is 50' but another value from 35' to 60' can be selected by the applicant without an ELOS. For a normal touchdown, consistent with non steep approach landings, the rate of descent should not exceed 6 fps. Although 6 fps may appear high for a business jet, in the 680 family at Maximum Landing Weight (MLW) or below, the maximum touchdown rate is 10 fps. Anything above this requires a hard landing inspection.

Of more interest to this paper are the abused conditions that need to be demonstrated and this is where EASA differs significantly from the FAA. EASA states that it must be possible to achieve an approach path angle 2 deg steeper (in Lugano's case this would be 8.0 deg) than the selected approach path angle in all configurations which exist down to the initiation of the flare, which must not occur above 150 % of the flare height. This allows Textron Aviation to demonstrate the abused conditions at 52.5' vice 35'. The FAA in paragraph 231d(1)(a) requires under calm air conditions to demonstrate that it is possible to complete an approach, touchdown, and stop without displaying any hazardous characteristics on an approach path angle 2 deg steeper than the steepest approach path angle. For this condition the rate of descent should be reduced to no more than 3 fps at touchdown and it needs to be demonstrated at the applicant's selected flare height. To put the FAA requirement in perspective, on an 8.0 deg path the descent rate is approximately 1,550 ft/min allowing the pilot 1.4s to arrest the descent rate when the flare is initiated at 35'.

When the legacy Sovereign was certified in 2006, the FAA guidance was AC25-7B which does differs from AC25-7C. For instance, when an approved means of approach path visual guidance is used, the landing flare height can be reduced to 35'; i.e. no ELOS required for a lower flare height. However, the most significant difference is in the flare height for the +2 deg abused approach. AC25-7B did not require the abused condition to be flown to the applicant's selected flare height which in the Sovereign's case was 35'. Instead the +2 deg abused condition was demonstrated at 150% or 52.5' which is consistent with current and past EASA regulations.

As mentioned earlier, the steep approach configuration for the 680 family is unique and therefore requires a significant amount of Sub Part B testing. For efficiency, Textron Aviation usually conducts this testing while completing the required testing for initial TC. This Sub Part B testing includes multiengine landing climbs, maneuver margin, multiengine lateral control, static longitudinal stability, static lateral directional stability and stall speeds and characteristics. Since FIKI was also desired for steep approach, some ice shape testing was also required with holding and pre activation shapes. Holding ice shapes represent 45 minutes of collected ice on the unprotected surfaces

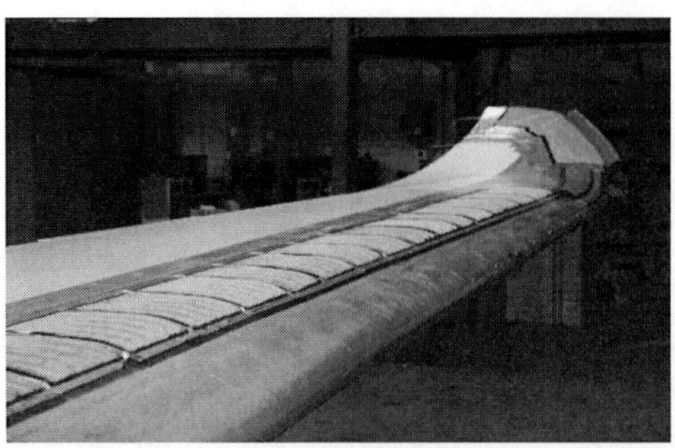

Figure 6 - Runback Ice Shapes

with all the protected surfaces clean and runback shapes as shown in Figure 6. Pre activation ice shapes represent the initial ice accumulation prior to the anti ice system fully operating. Pre activation ice is typically called sand paper ice and can be seen in Figure 7.

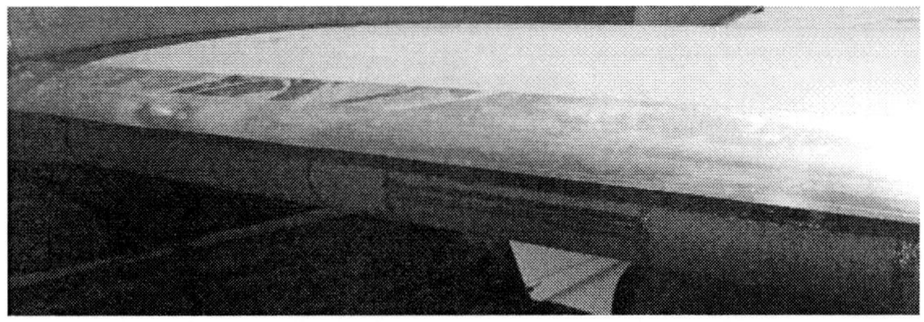

Figure 7 - Sandpaper Ice

In addition to the general ice shapes and Sub Part B testing, there are also some specific tests for steep approach. These tests require normal steep approach landing distance which is flown to 4.5 and 5.5 deg (with the steeper angles done by analysis), autopilot and flight director performance on the steeper beam and engine failures. The intent was not to certify the Latitude for single engine steep approaches. However an engine failure at decision height continuing to a landing and to a go around still needed to be demonstrated for compliance. The final steep approach testing was the abused testing at the steeper approach angle of +2 deg.

UNCONTAMINATED DEVELOPMENT PROGRAM

As with all flight test programs, there were some schedule constraints. The prototype aircraft was configured with ice shapes for initial TC testing. The installation of ice shapes takes a significant amount of time and resources. It was decided that collecting the specific steep approach test conditions while the prototype was already configured made sense even though steep approach certification was a post TC project. However, we had not collected any uncontaminated wing data for steep approach so the team needed to get that initial development completed in order to open up the envelope for ice shapes.

Going into this initial development program there were some important considerations, both positive and negative. First, the legacy Sovereign was previously certified for steep approach and the Sub Part B results for the 680 family of aircraft were very similar.

However, steep approach with FIKI was a new capability and the regulatory requirements had changed requiring abused testing to be shown at the flare height instead of 150% of the flare height. The initial development plan was to collect data to satisfy both the FAA and EASA requirements since it was unlikely the aircraft could successfully demonstrate an 8 deg abused approach down to 35′, but should be able to for 52.5′.

Thirty runs were spread over two flights to collect the required data. Flare height was 52.5′ and then 35′. Weight was +/- 10% of MLW and the approach angle was built up in 0.5 deg increments from 4.5 deg to 8.0 deg. As is typical at Textron Aviation, two different crews composed of three pilots total flew the runs. This allowed some concurrent training for the two pilots that had not flown steep approach testing before. Each pilot flew approximately ten runs at various flare height and approach angle combinations. Some touchdown rates at the lower flare height were high at 8.9 fps. However, these firmer touchdowns were correlated with a lazy pull or a late flare and were repeated as necessary to demonstrate satisfactory results. The initial clearance that came out of this testing was as follows: normal steep approach acceptable at 4.5 and 5.5 deg with a flare height of 35′, abused 7.0 deg and below was acceptable for a 35′ flare height and abused 7.5 deg and below was acceptable for a 52.5′ flare height. As is typical for tests that require buildup, the harder (steeper and lower) conditions were done later in the day on both flights which resulted in lighter weights as fuel was burned even though the conditions were still completed in the MLW band.

ICE SHAPES DEVELOPMENT

With this initial uncontaminated wing clearance, ice shapes steep approach development testing commenced. The aircraft was configured with holding ice shapes and a new crew member was added in the right seat to the pilot mix. The plan was to build up in approach angle from 5.0 to 8.0 deg for a flare height of 52.5′ and from 5.0 to 7.0 deg for a flare height of 35′. During the first sortie, ten runs successfully demonstrated 8.0 deg at 52.5′ but there was significant scatter in the data for the 7.0 deg at 35′. Since the gross weight was approaching the bottom of the acceptable band, the crew took a break and added fuel to bring the aircraft back to the top of the MLW band. Three runs were completed targeting 7.0 deg and a 35′ flare height. The actual results were as follows: first run was an approach angle of 6.5 deg, flare height of 53′ and a touchdown of 6.2 fps, the second 7.4 deg, 52′ and 6.8 fps. The third and final run of the day was a 7.0 deg approach angle, flare at 41′ with a touchdown rate of 16.7 fps! Instead of making it the usual touch and go to set up for another run, the crew made it a full stop and headed back to the hangar to evaluate what happened.

GO FORWARD PLAN

Although the touchdown was very hard, no actual aircraft damage occurred. However, it took ten days to get a valuable test asset airborne again after all the engineering analysis and inspections. Initial data review, especially on the flare technique, did not reveal any obvious cause for the hard landing. It was therefore suspected that there must have been an unknown gross weight sensitivity that would need to be managed. In addition, the now required extra development testing would need to find a way to reduce the scatter in the data, specifically flare height. This would allow us to approach the cliff safely and consistently without going over the edge again.

The initial actions were to revisit the Test Hazard Analysis worksheets to see if any new mitigation strategies could be employed. Out of this discussion, the following go forward plan was implemented. First, an initial up and away investigation on aircraft response at varying approach angles was planned. However, when this was flown, it was not useful because it is not possible to determine differences of less than 5′ in the flare height up and away. The next step was to add a gross weight (GW) buildup of 1,000 to 1,500 lbs in addition to the descent angle buildup of 0.5 deg increments and flare height reduction from 52.5′ to 35′. A predetermined knock it off criteria was also added for touchdown rate which was well below the requirement for a hard landing inspection. Since the abused landing conditions required touchdowns less than 3 fps, we decided that consistent results around 6 fps or higher were enough to demonstrate that a particular flare height and approach angle combination was not going to be acceptable. Also when a run was successfully completed, it needed to be repeated multiple times (usually two, maybe three repeats) to ensure that consistent results for that GW, approach angle and flare height could be achieved. This prevented the team from having a "lucky" pass and thinking it was acceptable to move on. Finally, and probably most importantly, crew consistency was ensured. A single crew with only one pilot flying was planned to fly all the uncontaminated wing conditions. This created a consistent flare technique and evaluation of aircraft response and provided continuity when moving from one flare height approach angle combination to the next. With those changes in place, the test team deployed to Roswell, NM for 150+ runs of development flying.

Figure 8 - 7.5 deg Approach to KROW

RESULTS AND LESSONS LEARNED

The mitigation strategies employed were effective and the uncontaminated wing development was successfully completed with no hard landings. At a flare height of 52.5', an abused steep approach of 8.0 deg with a touchdown less than 3 fps was repeatable. At 35', 7.0 deg was the abused approach angle limit. Both of these cases were demonstrated to a gross weight slightly above MLW. With this steep approach envelope defined, the ice shapes testing occurred using a slightly smaller, but similar buildup plan and successfully cleared the same envelope. However during 150+ runs, a few lessons are definitely learned.

One of the biggest takeaways from this program was to ensure adequate build up for the crew, not just the aircraft. The incident occurred at a 7.0 deg approach angle with a flare height of 41', but during the subsequent development 7.0 deg was demonstrated all the way down to 35'. At lower aircraft weights, successful runs were completed at 7.5 deg at 35'. A gross weight effect did exist but contrary to what was initially suspected it was not a major contributor to the hard landing. The incident was caused by a lack of crew buildup. The team did not allow the crew to develop a consistent steep approach flare technique. Due to training, the aircraft buildup had been spread out over three different pilots, resulting in only about ten runs per pilot with multiple approach angles and flare heights. When required to execute the flare with less than a 5' tolerance to avoid the "cliff", ten runs was simply not sufficient.

An unsuccessful steep approach could result in major damage to the aircraft, so it was essential to have a quantitative metric to predict future success. In this case, flare height scatter was the key safety parameter. Developing a consistent flare height at the initial 52.5' allowed a direct measurement of sufficient crew buildup and once flare height could be repeated within 5', the team knew they could move to 35'. This not only applied to initial crew buildup but also to each new day of testing. Practice runs were always conducted first and once flare height scatter was within 5', the crew could pick up where they left off from the previous day.

Crew buildup doesn't just apply to the pilot flying. The pilot monitoring is a critical component of any test team and in the case of steep approach was responsible for building the cadence down to the flare. The right seat pilot provided a radar altimeter countdown while the pilot flying concentrated on airspeed and glidepath control waiting for the appropriate radar height call to initiate the flare. This critical interplay was made clear when the test team relocated back to Wichita to finish up a few runs and a different crew member was placed in the right seat. It was quickly discovered that the two right seat pilots provided a completely different radar altimeter countdown. After approximately five runs, the cadence was adjusted and flare height scatter was once again reduced to less than 5'. Had buildup runs not been conducted at 52.5' first, the results to the aircraft could have been severe.

Flare height scatter was also very useful in knowing when to cease testing for the day. Air quality was critical, especially at the 35' flare height. AC guidance requires calm air, but the critical component is smoothness. Calm winds with thermals later in the morning did not work. As air quality slowly degraded throughout the morning, its effect could be measured in the flare height scatter. Once that scatter approached 5', testing had to be concluded for the day. Buildup always places the easier runs first and leaves the hard conditions to the end when the air quality has inevitably degraded. There is always a lot of pressure to get the last few runs complete even with marginal air quality. However, this program had the "benefit" of knowing what would happen if we missed our flare by a few feet. Having the aircraft down for maintenance for another ten days was not acceptable, making the choice to call it a day fairly easy.

From the beginning of test pilot school, the importance of a good flight report is drilled into pilots. However there are times when words can't pass on the experience gained by actually executing the maneuvers. With steep approach, providing yoke displacement and force can't replace learning the appropriate muscle memory ensuring the flare is consistently executed. Additionally, having actual past experience allows the pilot to not flinch and miss gathering the required data. Our team knew there was a cliff out there, but we also knew that previous runs had been successful and we could trust our technique to get consistent results as we moved on. Seeing with your own eyes that you can be successful is far more convincing that reading in a flight report that someone else was successful. Finally, having that experience allows the pilot to develop a sight picture to judge margin. On some runs, the data stream looked great but it was obvious in the cockpit that there was no margin remaining. The opposite also occurred when it was clear that there was plenty of margin and we could move to the next condition without further buildup. These are all items really hard to pass on in a flight report, making it necessary to dedicate time to crew build up in order to be successful.

Finally, real time feedback is critical. In all the steep approach testing, telemetry was used so the test team could provide feedback on actual approach angle, flare height and touchdown rate for each run. This allowed the team to accurately evaluate the success of each run based on actual results not the targets. Although it seems obvious, you

Figure 9 - 8.0 deg Approach KICT

must always base your decision to move to the next point on what you actually flew and real time feedback is critical in achieving this.

CONCLUSION

Regulatory requirements can sometimes drive the team to test in a corner of the envelope that has little to no margin. Time and effort is usually focused on providing time to buildup and understand the aircraft's capabilities, but the crew also needs time for sufficient buildup. Finding a key safety parameter is especially useful in identifying when enough crew build up has occurred so they can safely approach the low margin test conditions and also when to move away from those test points due to degraded environmental conditions or other factors. With careful planning, dedicated crewing and real time feedback, test conditions with little to no margin can be successfully completed, moving a test program forward.

Scott Martin & Kevin Claffy

Gulfstream

G500 Flight 000001-18 May 2015 1438Z

Video Presentation in
Original e-media

Gulfstream

0+30

G500 Flight Test Program

- Program Objectives
- Overview of Design
- Engineering Design and Development
- Preparations for Flight Test
- Initial Flight Test Objectives – Phase 1
- Lessons Learned
- Way Forward

Gulfstream

G500 Program Objectives

- Expands the family of Gulfstream business jets
- All new, best-in-class aircraft that builds upon Gulfstream's innovative technology leadership
  - Class leading high-speed range
  - Environmentally friendly
  - Larger, more spacious cabin
  - Revolutionary flight deck
  - Designed w/ "Voice of customer"

- Configuration Commonality
- Common Type Rating
- Family Approach Used to Minimize Non-Recurring Cost, Recurring Cost, Product Support Cost and Time-to-Market Schedule

Gulfstream

Gulfstream Product Lines

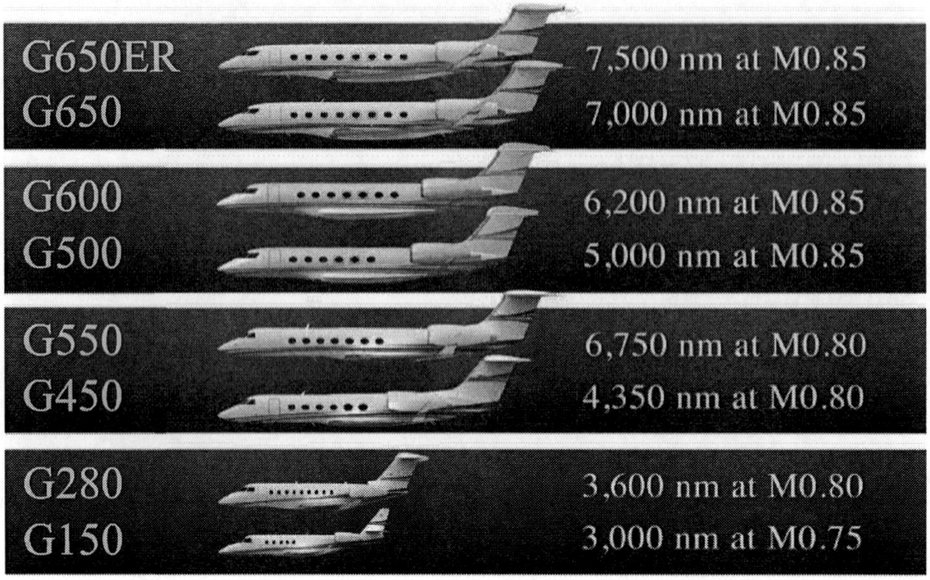

Gulfstream

G500 Overview of Design

- Overview of Design
  - Pratt & Whitney Engines
  - Digital Fly-by-Wire Flight Controls
  - Active Control Sidesticks
  - Touchscreen Controls & Indicators
  - Data Concentration Network
  - Systems Integration and Automation

Gulfstream

G500 Overview of Design – Suppliers

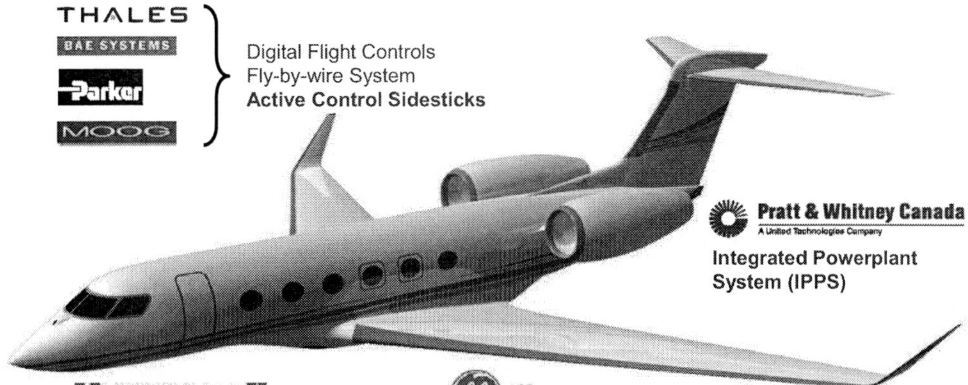

THALES
BAE SYSTEMS
Parker
MOOG

Digital Flight Controls
Fly-by-wire System
Active Control Sidesticks

Pratt & Whitney Canada
A United Technologies Company

Integrated Powerplant System (IPPS)

Honeywell

Avionics
Touchscreen Controllers (TSC)*
APU
ECS
Cabin Pressure Control

GE Aviation

Electrical Power Distribution
Aircraft Health & Trend Monitoring
Data Concentration Network

Gulfstream

G500 Cockpit Design Philosophy

- **Cockpit Re-design**
  - **By pilots, for pilots**
  - **Built-in Cockpit Flows**
  - **Touchscreens Functionality**
  - **Phase of Flight Operations**
- **Engines**
  - **Pure Power PW800 Series**
- **Flight Controls**
  - **Digital Fly-by-Wire**
  - **Active Control Sidesticks**
- **Evolution of G650**
  - **Avionics Functions**
  - **Cabin Essential**

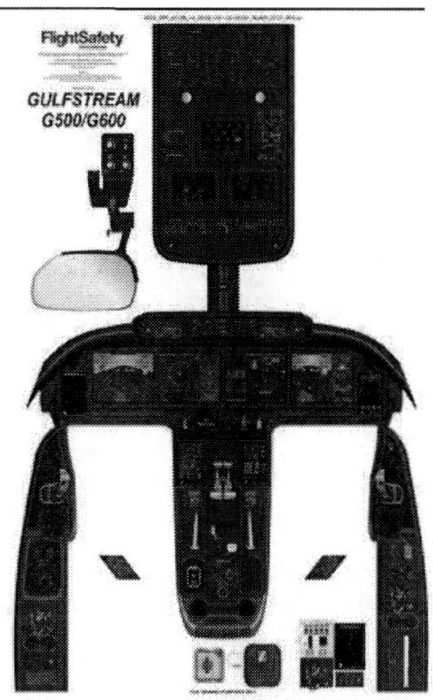

Gulfstream

G500 Digital Fly-by-Wire Flight Controls

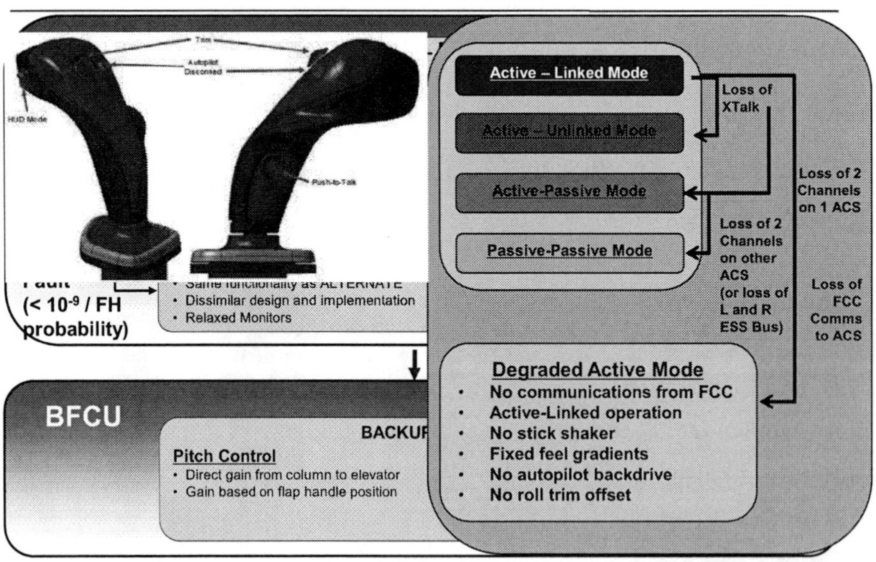

Gulfstream

G500 Active Control Sidesticks

- Left and Right Unique P/Ns
- Electrically Linked "Backdrive"
- Integrated Functions
  - Pitch/Roll Trim
  - AP Disconnect / Trim Reset
  - Push-to-Talk / HUD / EVS

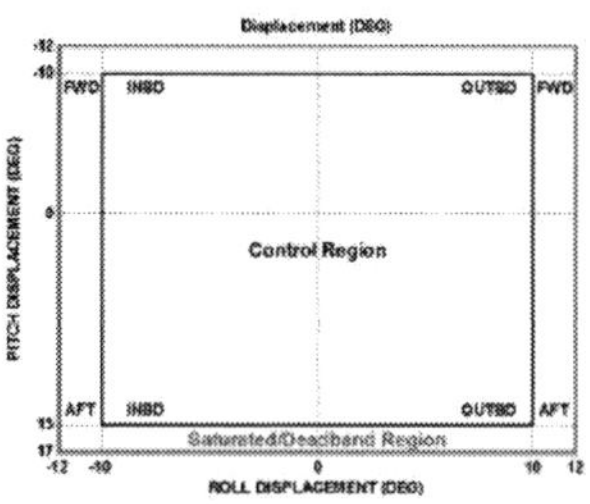

Gulfstream

G500 Active Control Sidesticks

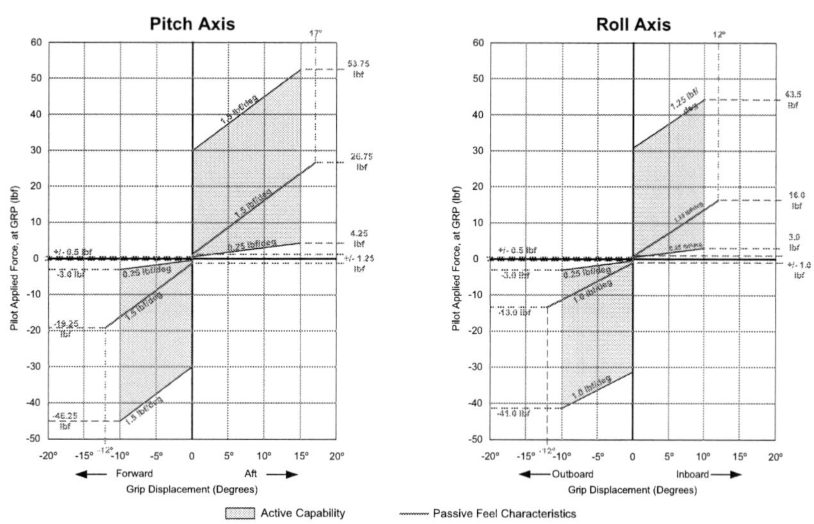

Gulfstream

G500 Active Control Sidesticks

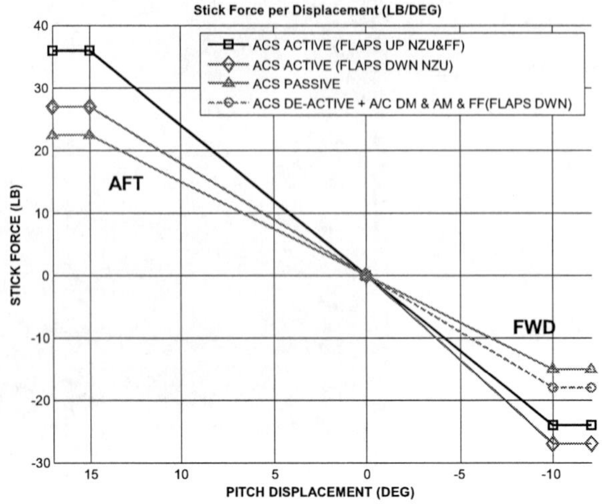

Gulfstream

G500 Active Control Sidesticks

- Insert Active Control Sidesticks video clip here…

Video Presentation in
Original e-media

Gulfstream

0+30

G500 Touchscreen Controls & Indicators

- Touchscreen Controllers (TSC)
- Overhead Panel Touchscreens (OHPTS)
- Standby Flight Displays (SFD)

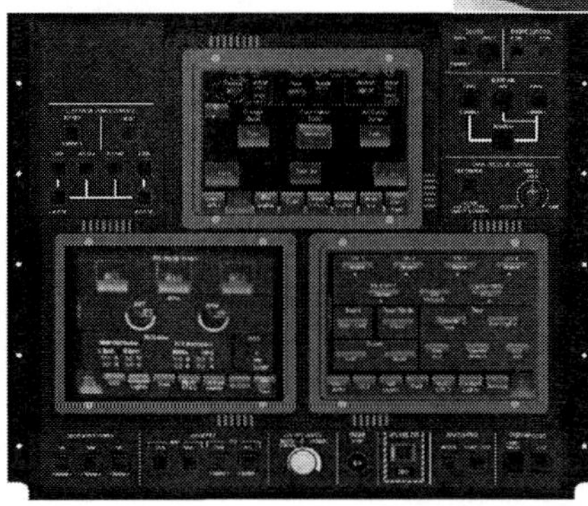

Gulfstream

G500 Touchscreen Controls & Indicators

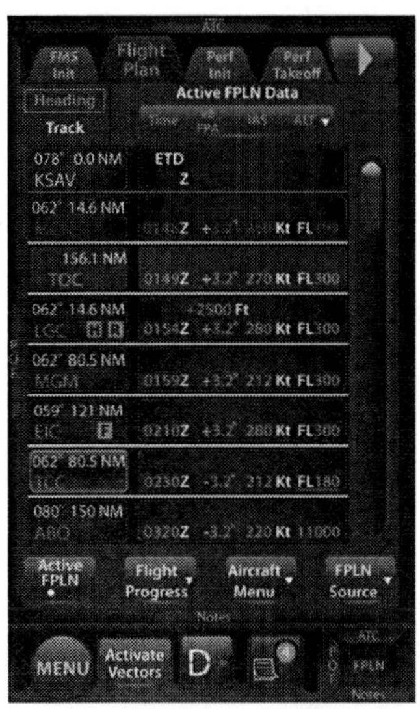

Gulfstream

G500 Touch User Interface

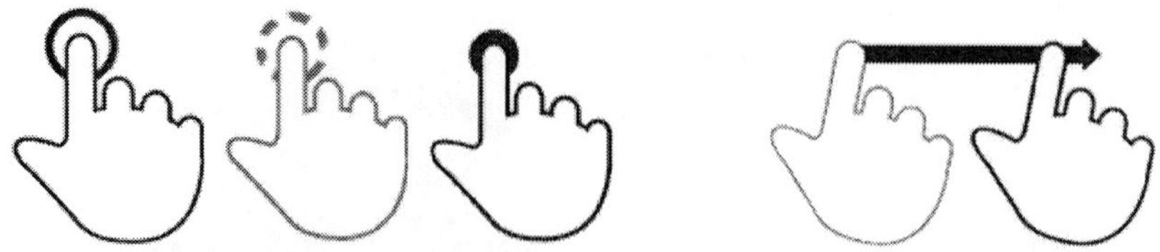

Land On Lift Off Press&Hold Drag or Scroll

Land On Location	Lift Off Location	
	On Target	Off Target
On Target	Activation	No Activation
Off Target	No Activation	No Activation

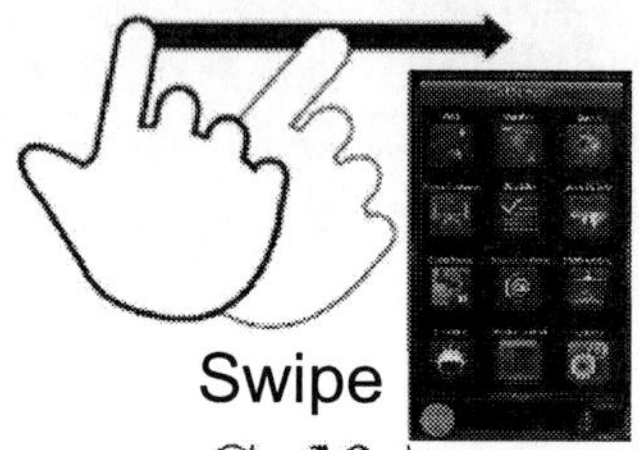

Swipe

Gulfstream

G500 Touchscreen Controls & Indicators

- Insert Touchscreen video clip here…

Video Presentation in
Original e-media

Gulfstream

0+40

G500 Data Concentration Network

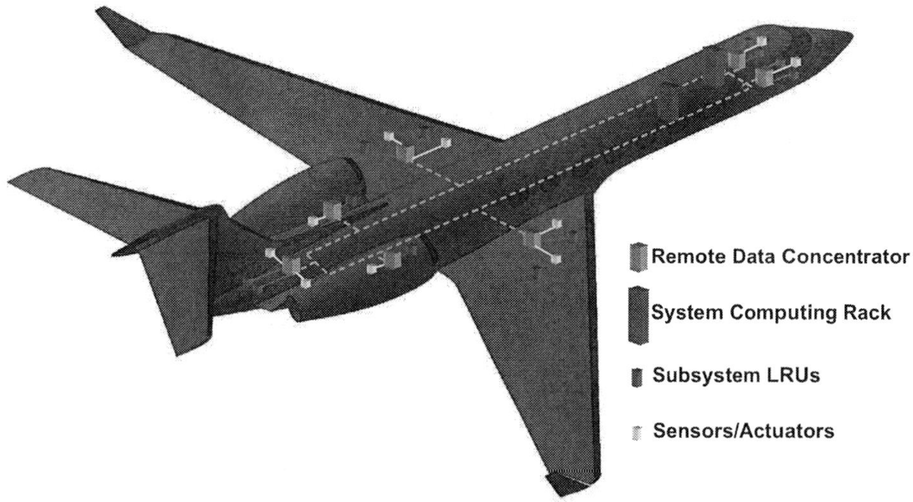

Gulfstream

G500 Systems Integration and Automation

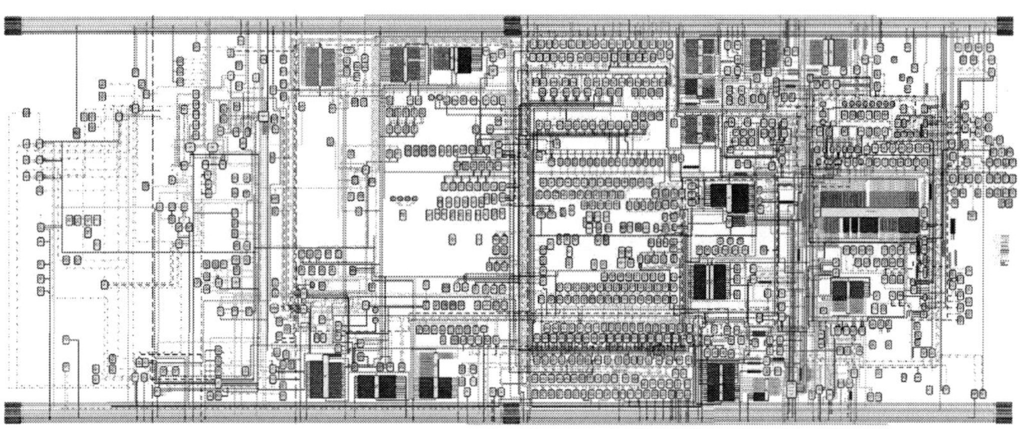

Gulfstream

G500 Engineering Design & Development

- Engineering Design and Development
  - Systems Integration Test
  - Ground Based Lab Simulation
  - Test Pilot and Engineering

Gulfstream

G500 Engineering Design & Development

- Insert Gulfstream Labs video clip here…

Video Presentation in
Original e-media

Gulfstream

1+15

G500 Engineering Design & Development

Gulfstream

G500 Engineering Design & Development

Gulfstream

G500 Engineering Design & Development

Gulfstream

G500 Engineering Design & Development

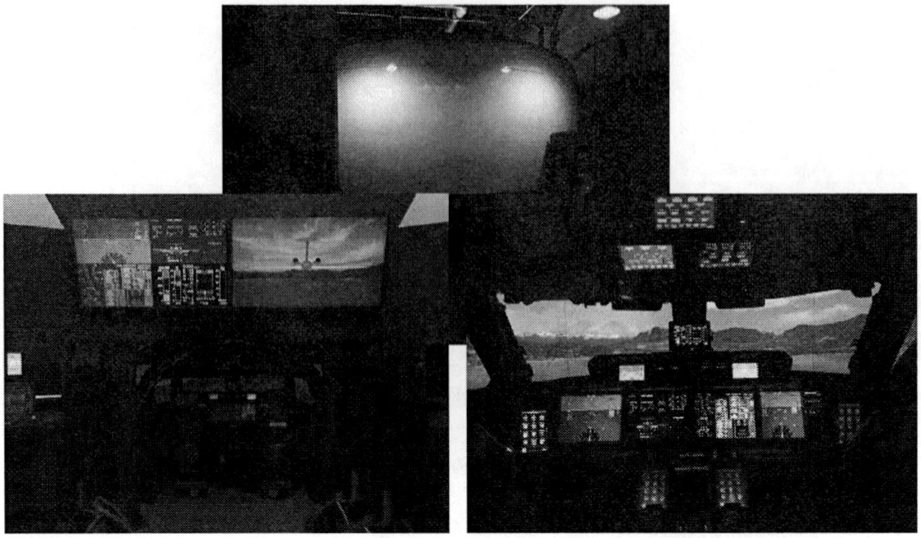

Gulfstream

G500 Engineering Design & Development

- ONERA F1 Wind Tunnel, Le Fauga, France
- Lift Characteristics in Ground Effect
- Affects of Thrust Asymmetry and Sideslip
- Influences of Aileron/Spoiler and Rudder Deflections

Gulfstream

G500 Preparations for Flight Test

- Preparations for Flight Test
  - Ground Based Simulation
  - Outside Operations
  - Systems Gauntlet
  - Low and High Speed Taxi

Gulfstream

G500 Initial Flight Test Objectives – Phase 1

- Initial Flight Test Objectives – Phase 1
  - Stage 1 primary objectives:
    - Handling qualities evaluation
    - Structural vibration
    - Engine operation
    - Systems operation

  - Stage 1 secondary objectives:
    - Flight test instrumentation
    - Air data system calibration
    - Suitability for envelope expansion

Gulfstream

G500 First Flight and Subsequent

- Test Objectives – Flight 1
  - Takeoff and Land
- Test Objectives – Flight 2
  - Fly again soon
  - Demonstrate normal cruise flight
  - Expand the flight envelope
- Test Objectives – Flight 3, 4, 5
  - Clear full stage 1 flight envelope
  - Demonstrate single failures of critical systems (flight controls, hydraulics, electrics)
  - Perform air starts of main engines and APU.
  - Flutter clearance for landing gear doors and fairings

Gulfstream

G500 Best Practices

- Best Practices
  - Pilots fully integrated into design and development
  - Investments in high fidelity laboratories
  - Crew Resource Management
  - Preparations for Test
  - Ground Operations
  - Flight Crew Training

Gulfstream

G500 Lessons Learned

- Lessons Learned
  - Develop Lab Simulation Early
  - Systems Integration and Automation
    - Digital Flight Controls & Active Control Sidesticks
    - Touchscreens – TUI / Color Philosophy
    - Airplane vs. Lab Simulation

Gulfstream

G500 Way Forward

- Accomplishments
  - S/N 72001
  - First Flight May 18th
  - 5 Flights in 12 days
  - 38,500 ft MSL
  - 250 KCAS / 0.80 M
- Test Program to Date
  - ~ 20 Flights
  - ~ 75 Hours
  - Envelope Expansion began August 31st
- Challenges
  - Flight Test Instrumentation

Gulfstream

G500 Flight 001-18 May 2015 Takeoff 1439Z

Video Presentation in
Original e-media

Gulfstream

1+20

G500 Flight 001 Aircrew and Ground Crew

Gulfstream

Questions?

Gulfstream

Modernizing the Safety Process:

Complexity Theory for Flight Test Professionals

Dan "Animal" Javorsek

Air Force Test Center

Edwards AFB, CA

Integrity - Service - Excellence

Motivation

- Fatal Mishap Report Observations
  - Deterministic narrative
  - Retrospective predictability
  - Rarely consider human agency
- Can other fields of study help us?...*Yes!*
  - Challenge assumptions
  - Address inconvenient truths
  - Invest in preparedness

Brief History of Fatal Mishaps

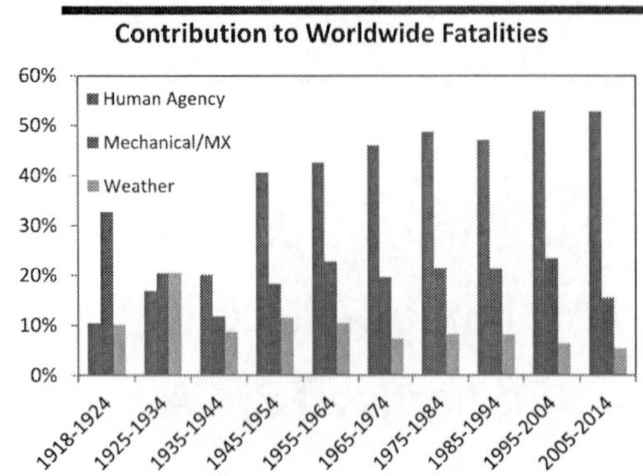

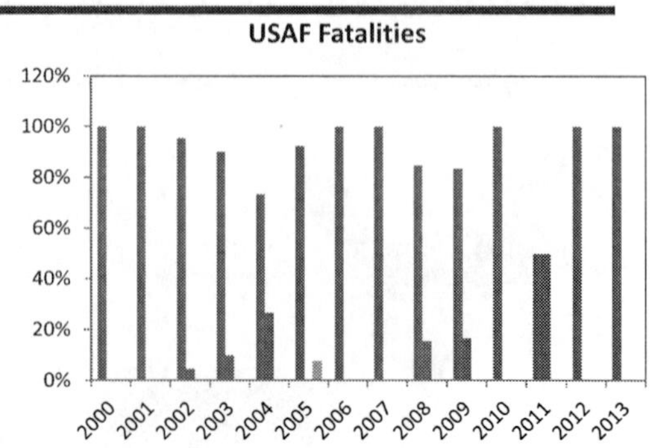

- Human agency as root cause in AFMC mishaps
  - 60% of fatal accidents
  - 80% of fatalities
- Data from Bureau of Aircraft Accidents Archives (BAAA) and USAF JAG Accident Investigation Boards (AIB)

3

Breaking Through the Plateau

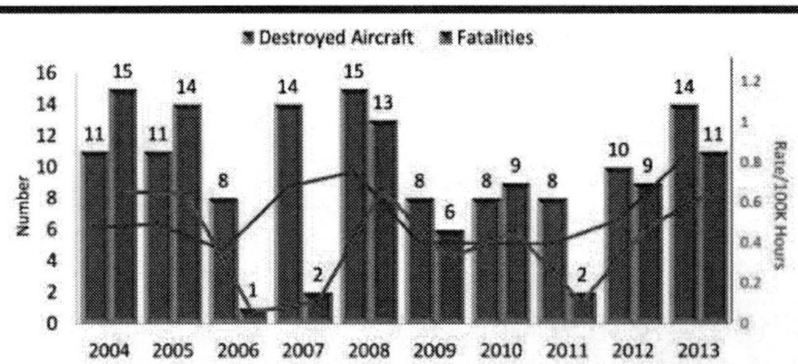

"While the last five years show continued improvement, the trending data illuminate a new concern – we've plateaued."

– MGen Woodward, Air Force Chief of Safety in reference to the $8.8 billion and 767 lives lost from 2002-2012

Change our approach to Human Agency

4

Complexity Theory 101

- Complex Adaptive System is made of agents that are:
  - Adaptable
  - Heterogeneous
  - Interdependent
  - Highly connected

- Used for systems that lack a clear set of governing equations
  - Embraced by economics, medicine, meteorology, ecology
  - Characterized by emergent macroscopic features
  - In such systems "control" is replaced by "influence"
  - Uncertainties can have disproportionately large impact

- Air vehicle system with human in the loop is complex

5

Complex Adaptive Systems

Complicated

- Example: Airplane
- Predictable
- Not as prone to surprise
- Distribution: Gaussian (bell curve)

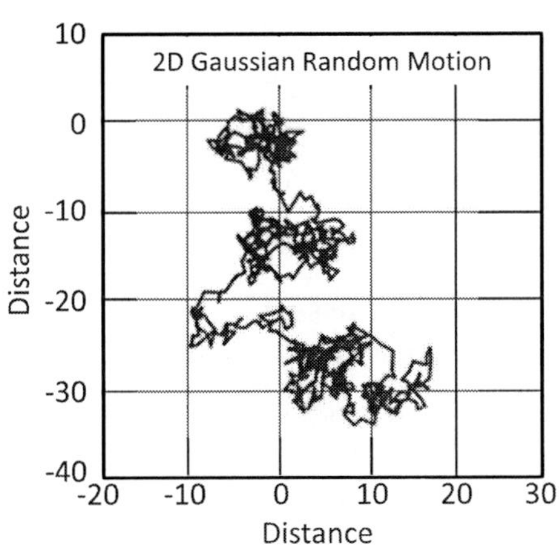

Complex

- Example: Air traffic
- Can appear predictable
- Susceptible to great surprise
- Distribution: Heavy Tailed

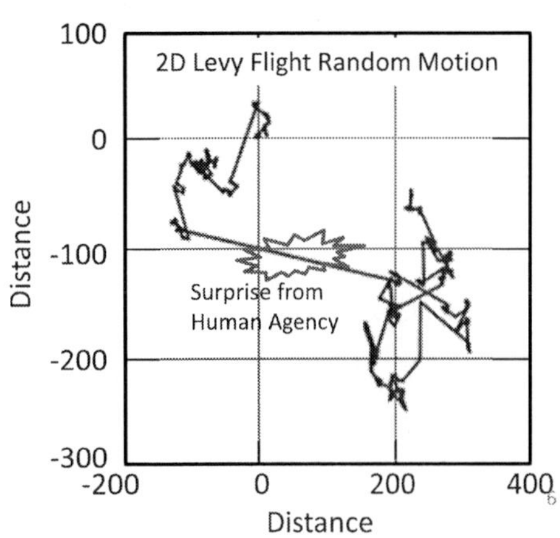

6

Cognitive Biases Increase Unpredictability

- Errors in judgment and decision making occur:
  - Subconsciously
  - Independent of demographics (age, income, education, etc.)
  - Due to neurological limitations
  - Due to limited capacity for info processing

- Lead to...
  - Perceptual distortion
  - Inaccurate judgment
  - Illogical interpretation

Bias	Description
Anchoring/Focalism	Rely too heavily on one piece of info
Automation Bias	Overdependence on automation
Base rate/Conjunction Fallacy	Focus on specific and ignore general info
Expectation/ Confirmation Bias	Focus on preconceptions
Halo Effect	Overall impression underestimates errors
Overconfidence Effect	Underestimation of uncertainty
Regressive Bias	Under estimation of low likelihoods and vv
Retrospective Distortion	Unpredictable event was preventable

Risk Assessment of Human Influenced Hazards

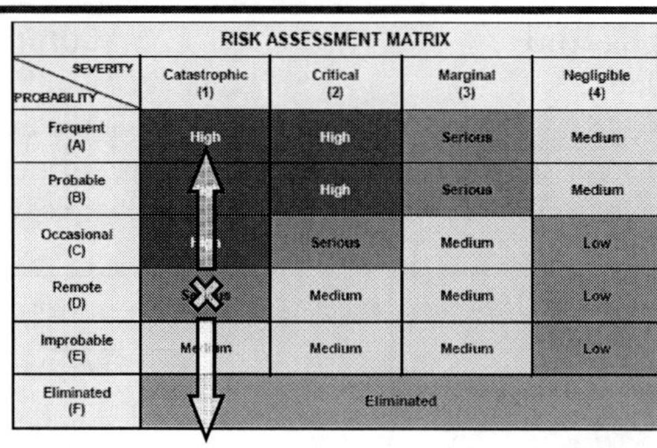

Pros

- Easy to use
- Widely distributed
- Responsible for reduction in absolute number of fatalities

Cons

- Probability assessment relies on "engineering judgment"
- Human agency treated same as other factors
- Assumes linear severity

8

Partitioned Multiobjective Risk Method

- Multiobjective problem
  - Severity levels of aircrew injury are not linear
  - Unable to treat cost and aircrew injury the same
  - Cost can use typical Bayesian expected value

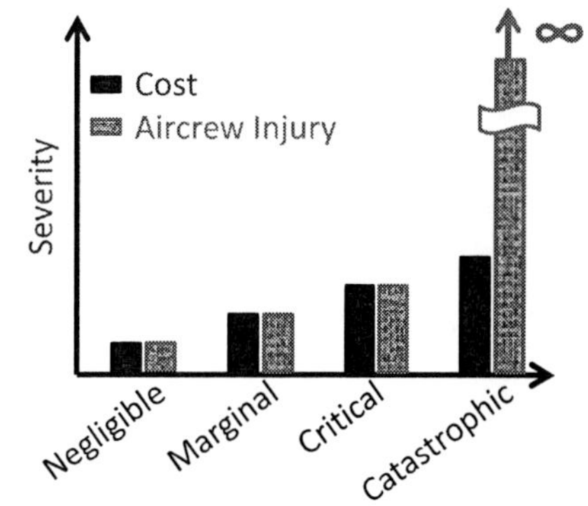

- Traditional expected value is inadequate for catastrophic events
  - Unable to differentiate between extremes
  - Fatal catastrophic loss must be handled differently from non-fatal incidents

9

Recommendations

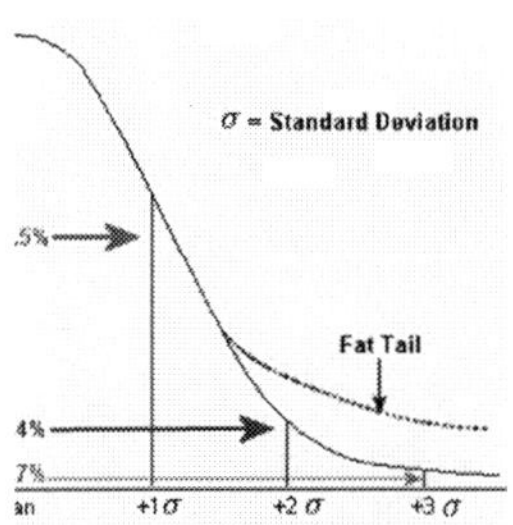

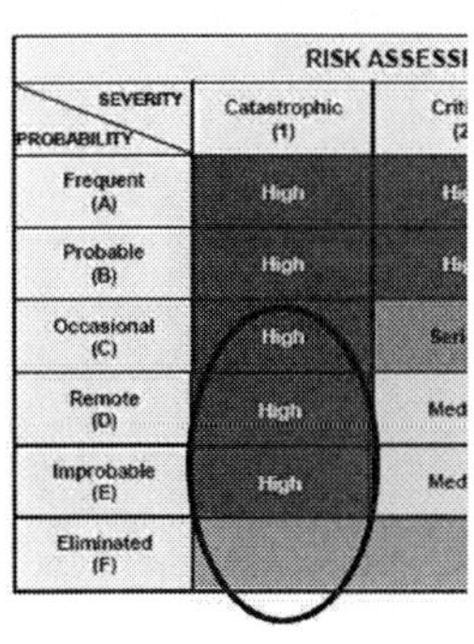

- Shift culture away from determinism
  - Anticipate and plan for human agency
  - Teach non-Gaussian statistics
  - Teach psychology of judgment & decision making
- Modify risk assessment for catastrophic events
  - Remove dependence on probability uncertainty
  - Employ the Partitioned Multiobjective Risk Method
- Continue to research and investigate complexity applications

10

Takeaways

- Be the assertive skeptic
- Nudge at every opportunity
- Remain vigilant for cognitive biases
- Plan for inevitable human agency

Questions?

12

Acknowledgments

- Special thanks to the following individuals who have helped focus this effort through their engaging discussions and exhaustive feedback:
  - Steve Rainey
  - John Schwitz
  - Brad Carlson
  - Mike Nielsen

13

Recommended Reading

- Flight Test Mishaps
  - Soucie, Cheek, *Safer Skies*, (Skyhorse Publishing:, 7 Feb 2015).
  - Merlin, Bendrick, Holland, *Breaking the Mishap Chain*, (Military Bookshop: 23 May 2012).
  - Gibb, Ercoline, Scharff, "Spatial Disorientation: Decades of Pilot Fatalities," *Aviation, Space, and Environmental Medicine* **82** (2011) 1-8.
  - Walters, Sumwalt, *Aircraft Accident Analysis: Final Reports*, (McGraw-Hill Education: 26 Jan 2000).
- Complexity Theory
  - Sole, *Phase Transitions*, (Princeton: Princeton University Press:,14 Aug 2011).
  - Sheffer, et al, "Early-warning Signals for Critical Transitions," *Nature* **461** (2009) 53-59.
  - Johnson, *Simply Complexity: A Clear Guide to Complexity Theory* (Oxford: University of Chicago Press, 2007).
- Statistics/Uncertainty/Psychology
  - Taleb, *The Black Swan* (New York, NY: Random House Trade Paperback, 11 May 2010).
  - Haimes, *Risk Modeling, Assessment, and Management* (Hoboken, NJ: John Wiley, & Sons Inc., 2009).
  - Mlodinow, *The Drunkard's Walk* (Vintage, 5 May 2009).
  - Tversky, Kahneman, "Judgment under Uncertainty: Heuristics and Biases," *Science* **185** (1974) 1124-1131

14

Database Information

- Worldwide data compiled from the Bureau of Aircraft Accidents Archives

  http://www.baaa-acro.com

- USAF data compiled from the Judge Advocate General Legal Operations Agency Claims and Tort Litigation Accident Investigation Board Repository

  http://usaf.aib.law.af.mil

15

PMRM Example: Hypothetical First Flight

- Assumptions:
  - Contractor facility outside metropolitan area
  - Contractor facility flight time over populated areas = 1%
  - Remote location flight time over populated areas = 0%
  - Probability of accident = 1/10,000
  - Remote area cost (vehicle) = $75M
  - Populated area cost (civilian casualties+law suits+collateral damage+vehicle) = $500M
- Expected Loss Cost Assessment:
  - Remote Location = $75M/10,000
    = $7,500
  - Contractor Facility = 99%($75M/10,000)+1%($500M/10,000)
    = $7,425+$500 = $7,925
- PM Decision: Expected Loss << Move to Remote Location
- According to PMRM....NO!!! A populated area crash is unacceptable!

16

Maintaining Relevance In The Midst Of Dramatic Strategic Change

Mr. Dan Vanderhorst (F)
Mr. Don Sheesley (AF)

Technology Explosion
100
1960 (F-4)
1964 (A-7)
1970 (F-111)
2006 (F-35)
*Data from "NASA Study on Flight Software Complexity",
3

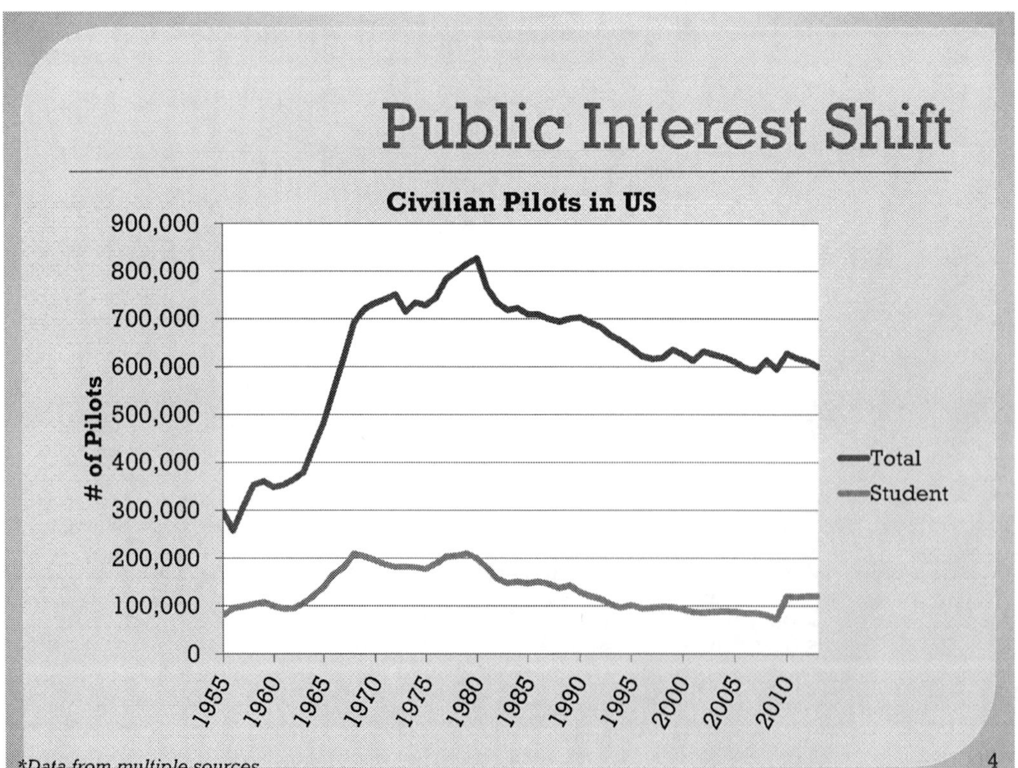
Public Interest Shift
Civilian Pilots in US
# of Pilots
900,000
800,000
700,000
600,000
500,000
400,000
300,000
200,000
100,000
0
1955
1960
1965
1970
1975
1980
1985
1990
1995
2000
2005
2010
Total
Student
*Data from multiple sources
4

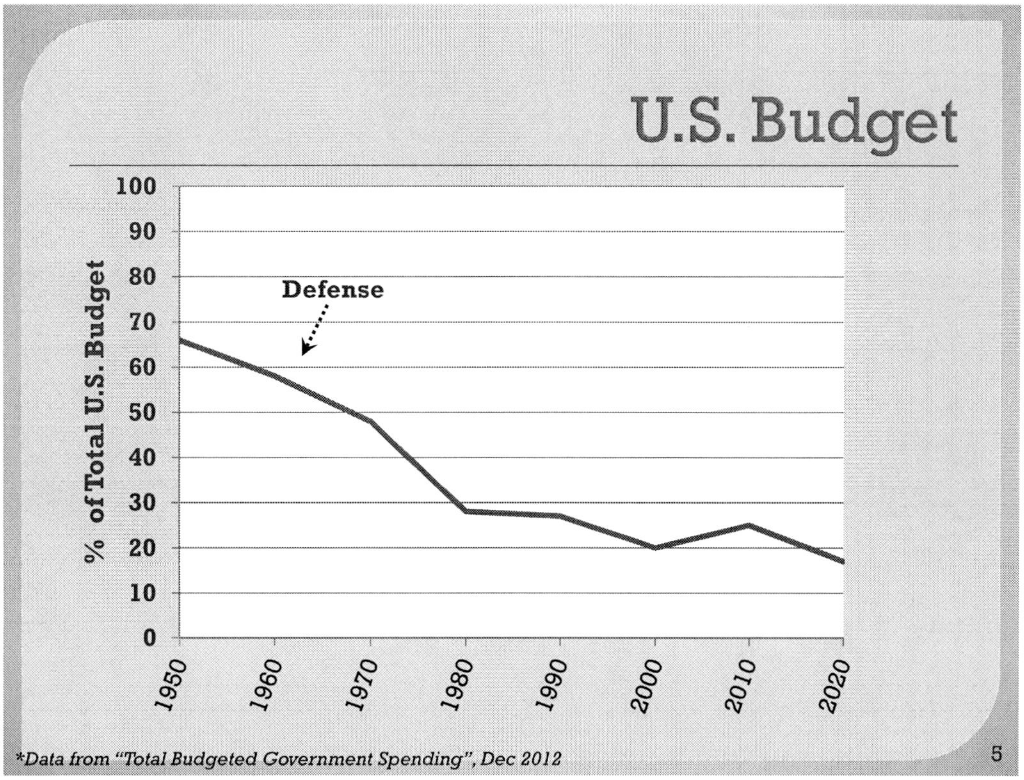
U.S. Budget
100
90
80
70
60
50
40
30
20
10
0
% of Total U.S. Budget
Defense
1950
1960
1970
1980
1990
2000
2010
2020
*Data from "Total Budgeted Government Spending", Dec 2012
5

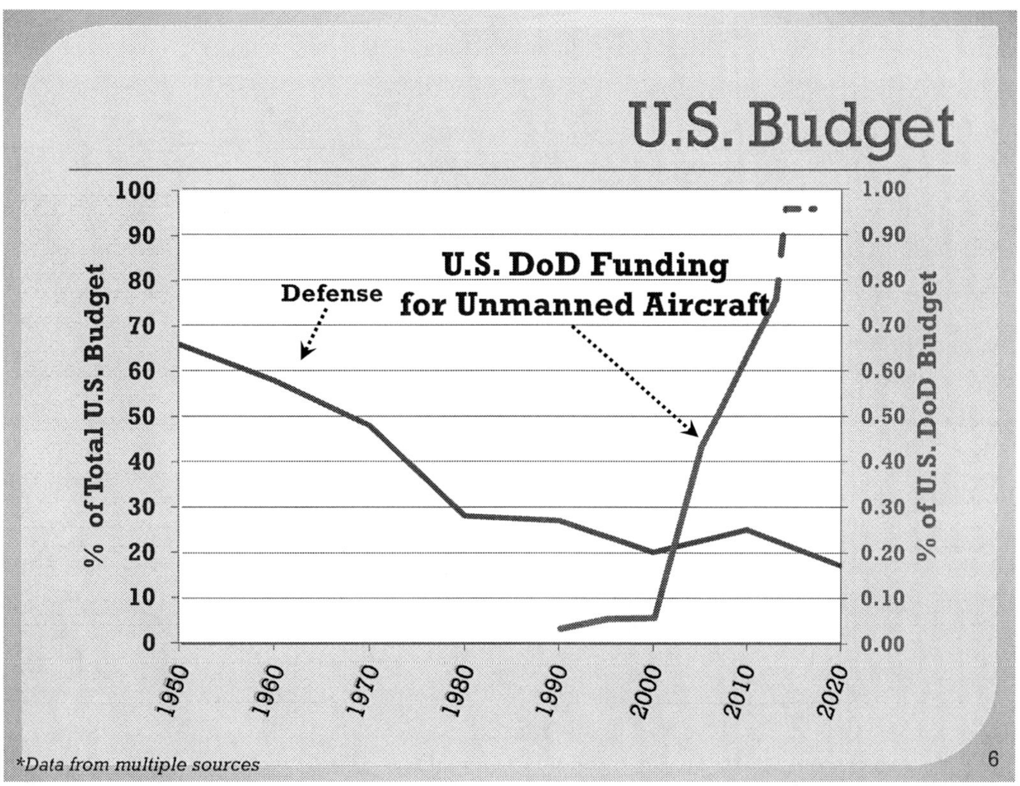
U.S. Budget
100
90
80
70
60
50
40
30
20
10
0
% of Total U.S. Budget
Defense
U.S. DoD Funding
for Unmanned Aircraft
1.00
0.90
0.80
0.70
0.60
0.50
0.40
0.30
0.20
0.10
0.00
% of U.S. DoD Budget
1950
1960
1970
1980
1990
2000
2010
2020
*Data from multiple sources
6

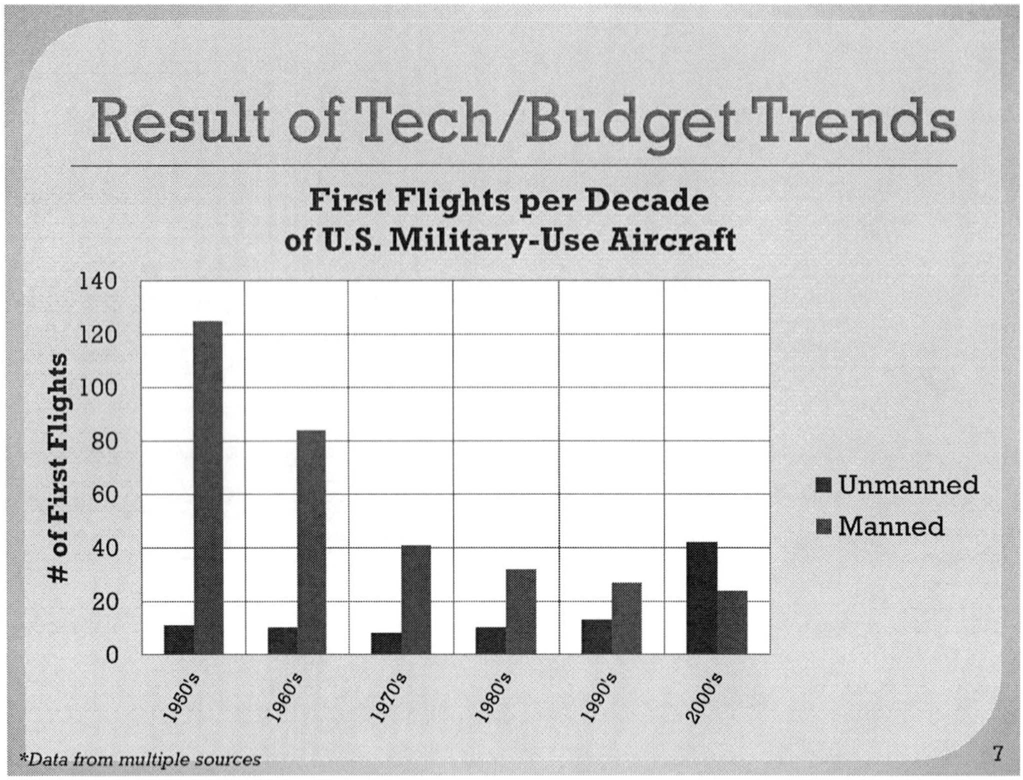
Result of Tech/Budget Trends
First Flights per Decade
of U.S. Military-Use Aircraft
140
120
100
80
60
40
20
0
# of First Flights
1950's
1960's
1970's
1980's
1990's
2000's
Unmanned
Manned
*Data from multiple sources
7

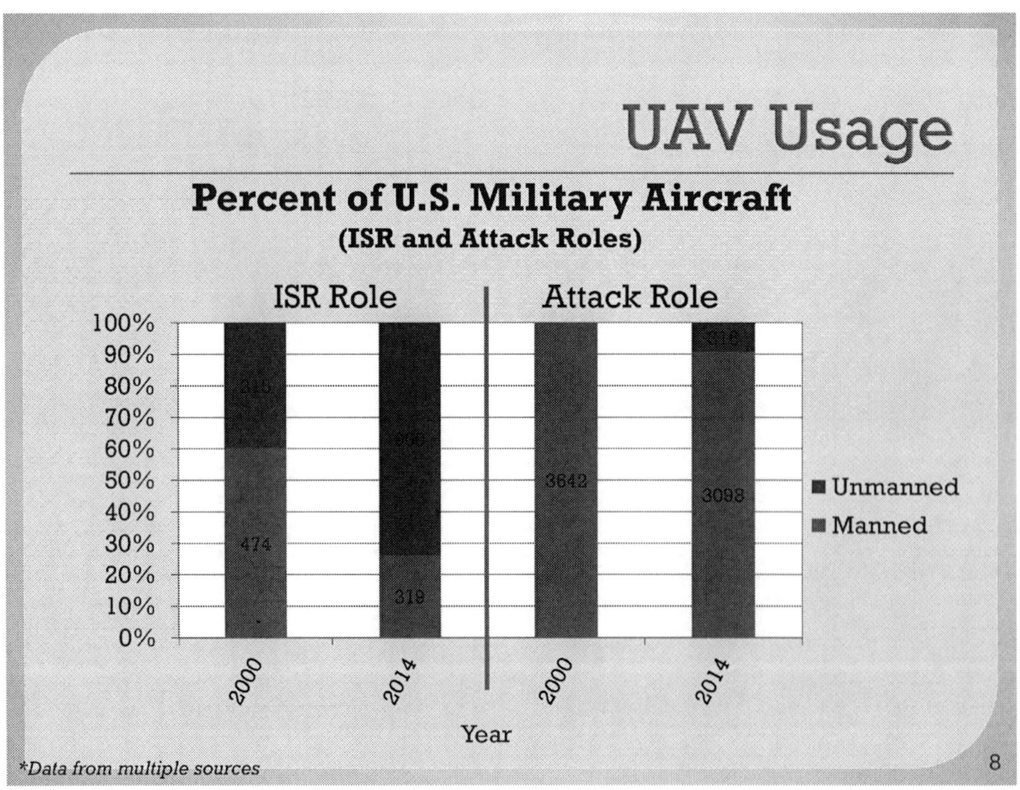
UAV Usage
Percent of U.S. Military Aircraft
(ISR and Attack Roles)
ISR Role
Attack Role
100%
90%
80%
70%
60%
50%
40%
30%
20%
10%
0%
474
319
3642
3098
Unmanned
Manned
2000
2014
2000
2014
Year
*Data from multiple sources
8

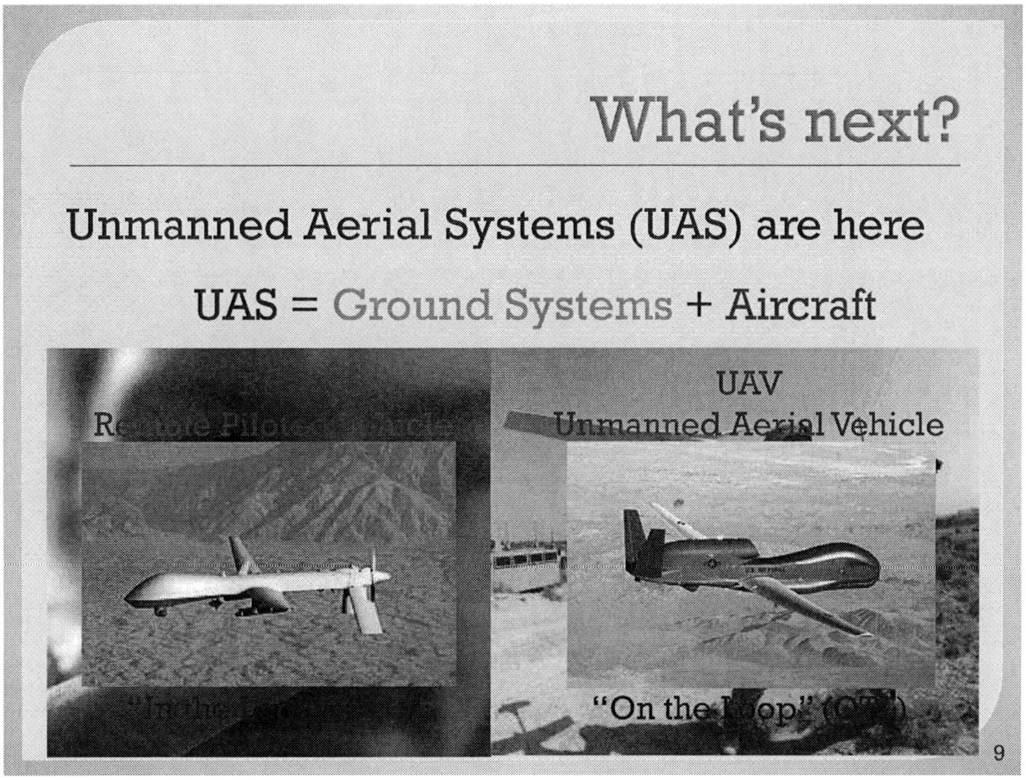
What's next?
Unmanned Aerial Systems (UAS) are here
UAS = Ground Systems + Aircraft
UAV
Unmanned Aerial Vehicle
9

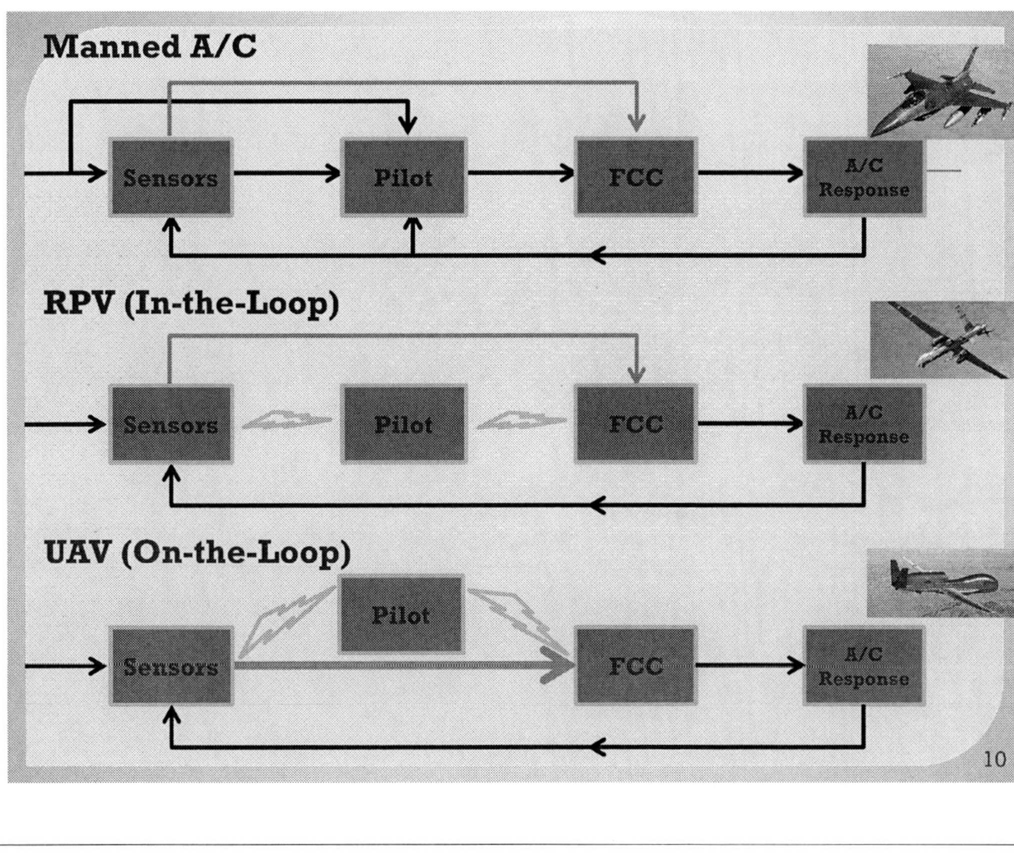
Manned A/C
Sensors
Pilot
FCC
A/C Response
RPV (In-the-Loop)
Sensors
Pilot
FCC
A/C Response
UAV (On-the-Loop)
Pilot
Sensors
FCC
A/C Response
10

Cockpit = "a compartment for the pilot"
11

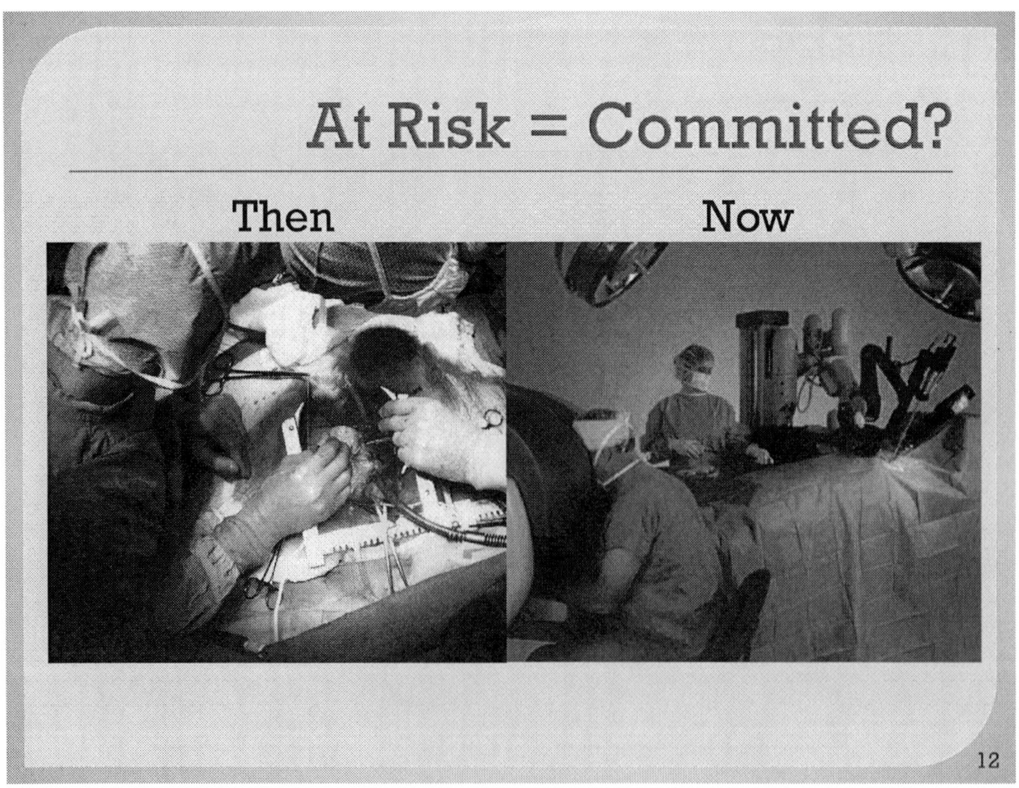
At Risk = Committed?
Then
Now
12

What's next?

Networked systems in control

13

SETP membership is strong!

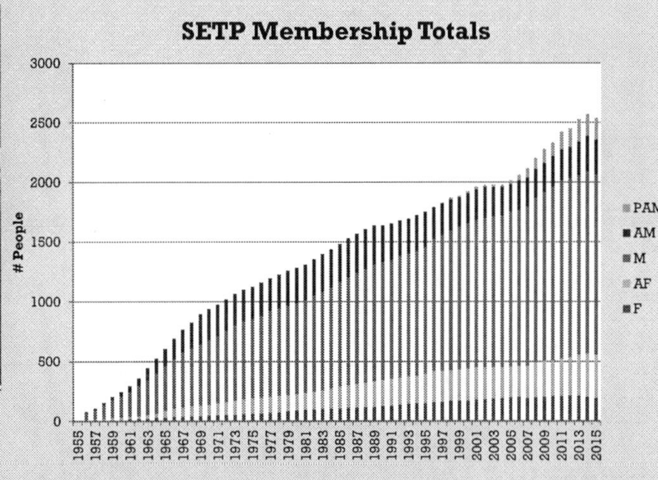

**Data from SETP membership files (thru Apr 2015)*

14

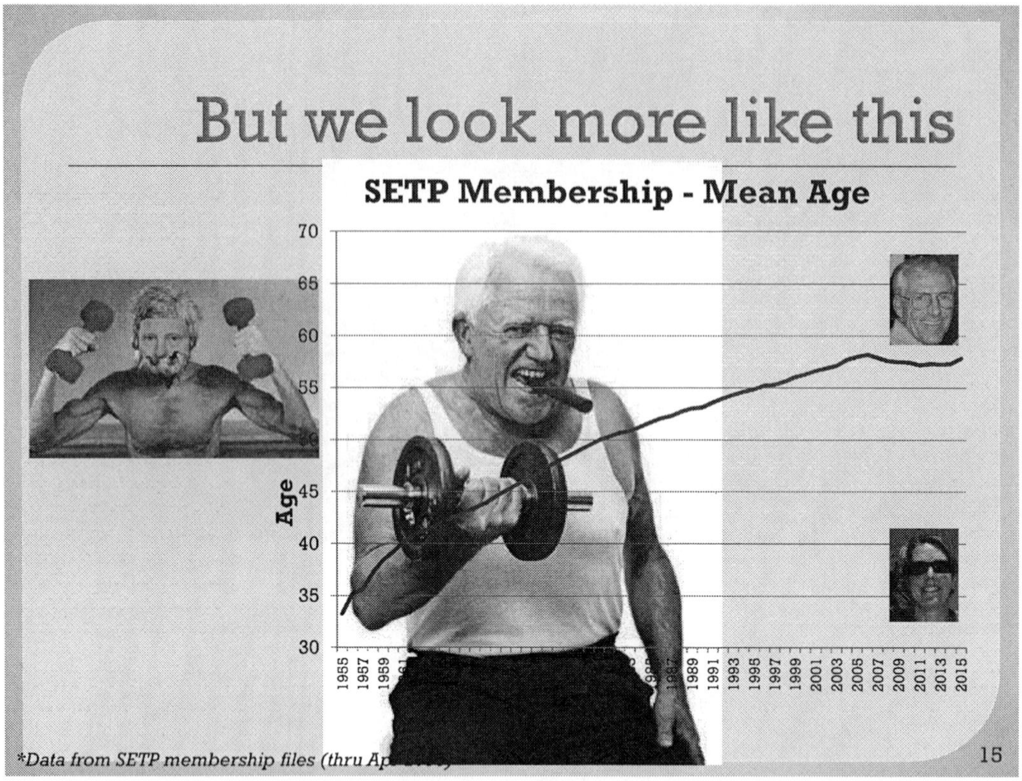
But we look more like this
SETP Membership - Mean Age
Age
70
65
60
55
45
40
35
30
1955
1957
1959
1989
1991
1993
1995
1997
1999
2001
2003
2005
2007
2009
2011
2013
2015
*Data from SETP membership files (thru Ap
15

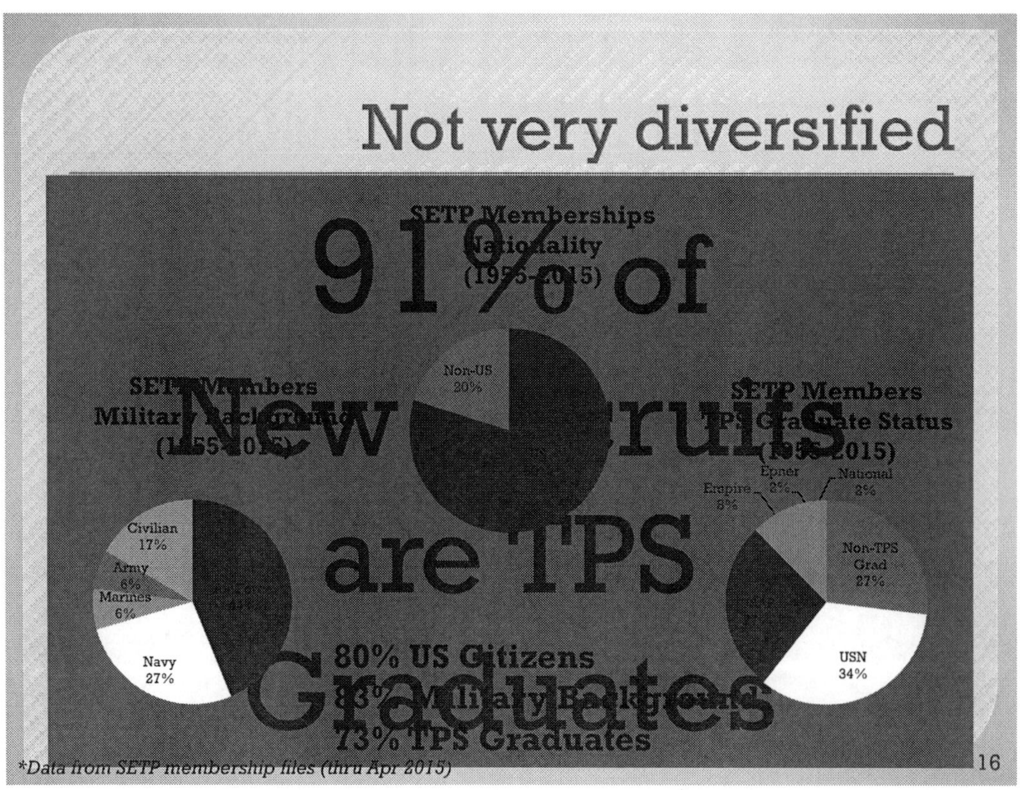
Not very diversified
SETP Memberships
Nationality
91% of
Non-US
20%
SETP Members
Military Background
New
ruits
SETP Members
TPS Graduate Status
(1955-2015)
Epner
2%
National
8%
Empire
8%
Civilian
17%
Army
6%
Marines
6%
are TPS
Non-TPS
Grad
27%
Navy
27%
80% US Citizens
USN
34%
Graduates
73% TPS Graduates
*Data from SETP membership files (thru Apr 2015)
16

How do we Maintain Relevance?

- Should we add a formal feedback loop to/from Test Pilot Schools?
- Should we do more educational presentations?
  - Flt Cntl, FQ, HQ testing, Operational testing
  - UAS testing, Network testing
- Should we update our membership criteria to include new & relevant areas of aerospace?
  - Ex: Certain types of UAS testing, Network testing

17

2012 SETP Committee led Paper

SETP UAV Development and Flight Test Committee

POSITION PAPER:

THE ROLE OF THE SOCIETY OF EXPERIMENTAL TEST PILOTS IN THE DEVELOPMENT AND FLIGHT TESTING OF UNMANNED AIR VEHICLES

Flight Testing Comparison

Subsystem	Manned	Unmanned	Change
Performance	Yes	Yes	None
Structural loads	Yes	Yes	None
Flutter	Yes	Yes	None
Propulsion	Yes	Yes	None
FQ/Aeromodeling	Yes	Yes	None
Handling Qualities	Yes	Yes	Different
Sensors	Yes	Yes	None
Datalinks (control)	Yes	Yes	None
Systems (incl. non-control datalinks)	Yes	Yes	None
Human Systems Integration/Human Factors	Yes	Yes	Different
Stores Cert	Maybe	Maybe	None
High AoA/Stall	Yes	Maybe	Some
Autoland and FMS	Maybe	Maybe	None
Ship Suitability	Maybe	Maybe	None
First Flight	Yes	Yes	None

18

2012 SETP Committee led Paper

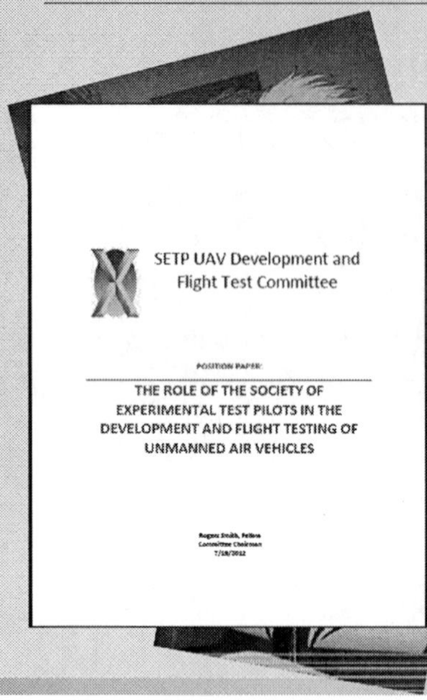
SETP UAV Development and Flight Test Committee

POSITION PAPER

THE ROLE OF THE SOCIETY OF EXPERIMENTAL TEST PILOTS IN THE DEVELOPMENT AND FLIGHT TESTING OF UNMANNED AIR VEHICLES

Subsystem	Manned	Unmanned	Change
Performance	Yes	Yes	None
Structural loads	Yes	Yes	None
Flutter	Yes	Yes	None
Propulsion	Yes	Yes	None
FQ/Aeromodeling	Yes	Yes	None
Handling Qualities	**Yes**	**Yes**	**Different**
Sensors	Yes	Yes	None
Datalinks (control)	Yes	Yes	None
Systems (incl. non-control datalinks)	Yes	Yes	None
Human Systems Integration/Human Factors	**Yes**	**Yes**	**Different**
Stores Cert	Maybe	Maybe	None
High AoA/Stall	**Yes**	**Maybe**	**Some**
Autoland and FMS	Maybe	Maybe	None
Ship Suitability	Maybe	Maybe	None
First Flight	Yes	Yes	None

19

What makes a "test pilot"?

Knowledge, Ability, Skill Sets of a Pilot

Experience, Judgment, Thought Process gained through Test Pilot Trng and/or Ops

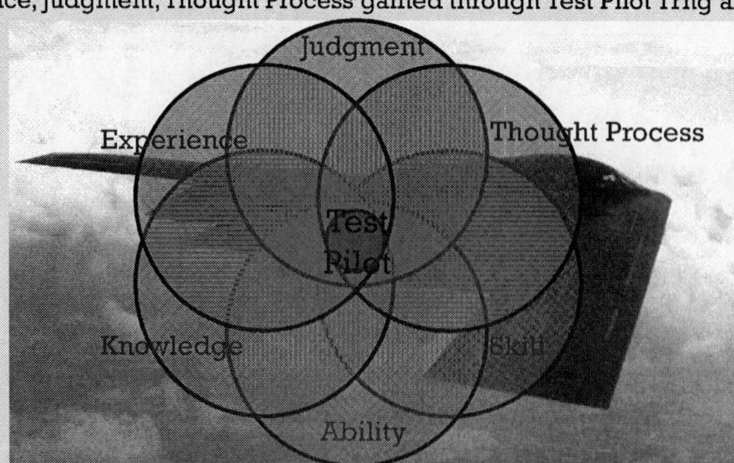

Question: Is Modern tech making it less about "Skill" and more about the other 5?

20

SETP Membership Counters

- Should some UAS tests be valid membership counters?
  - Co-pilots can be Associate Members right now
- AM Experience Requirements - Intentionally Low
  - TPS + 6 mo doing Engineering Evals
  - TPS + Astronaut **TRAINING**
  - (No TPS) - 6 Months as a Test Pilot and 6 Qualifying flights
- Bottom Line: Should SETP value the flight test experiences of people doing complex UAS testing?

21

Conclusions & Recommendations

Conclusions:

1. SETP is strong & very relevant to today's world of aviation.
2. Changes in the strategic environment will likely pose challenges to SETP maintaining relevance.

Recommendations:

1. Do more educational presentations and updates to SETP membership on what is being taught, latest techniques in testing Flt Controls, HQ, Systems
2. The Society should add formal feedback loops: SETP – TPS – SETP...focus on test technology, industry feedback, flight test techniques
3. Board of Directors reactivate UAV Committee to report on the potential incorporation of some UAS testing into SETP associate membership qualifications.
4. Members should discuss future directions for the Society targeting 5-10 years out.

22

Lessons Learned

- There is no provision in the Ray E. Tenhoff Award criteria for a non-technical paper
- Think critically about environmental changes
- Assess your contributions for relevance to today's & tomorrow's mission
- Come prepared to defend against offensive weapons or questions

23

References

- Images, *www.wikipedia.com*
- Global Movile Statistics. Research Statistics and Analysis. Mobi Forge.Mobi Web Services Division of AFILIAS. Dublin, Ireland. Web. Retrieved 2 March 2015
- "Software is steering Auto Industry" John Grapper. Financial Times, 18 February 2015. Web. Retrieved 2 March 2015
- "The Growing Importance of Sustaining Software for the DoD". Douglas C. Schmidt. Software Engineering Institute. Carnegie Mellon University. 1 August 2011. Web. Retrieved 3 March 2015.
- "General Information and Statistics, FAA Certificated Pilots." *www.aopa.org*. Retrieved 2 March 2015
- "Aviation Data Statistics, Certified US Pilots." *www.faa.gov*. Retrieved 2 March 2015
- "Resident Population Data – 2010 Census." *www.census.gov*. Retrieved 2 March 2015
- "Total Population of the United States 2004 to 2014." *www.census.gov*. Retrieved 13 March 2015
- "Software Engineer Salaries." *www.glassdoor.com*. Retrieved 2 March 2015
- "Pilot Salaries." *www.airline pilot central.com*. Retrieved 2 March 2015.
- "Military Budget of United States." *www.wikipedia.com*. Retrieved 2 March 2015
- "RoboEnomics." Tuesday January 14, 2014. Colin Lewis. Web. Retrieved 2 March 2015
- *www.f35.com*. Web. Retrieved 3 March 2015
- Chart Source Data: Cybertechuav.com, www.flightglobal.com, Armedforces-int.com, Defense-update.com, Tutorgigpedia.com, Selexgalieo.com, Robots.net, Defensereviewasia.com, Atuma.com.pk, Eckind.com, Zala.aero, Uav-group.dk, Flightinternational.com, Aerofiles.com, Wikipedia.com, Designation-systems.net, Driko.org
- "List of active United States Military Aircraft." *www.wikipedia.com*. Retrieved 21 April 2015
- GAO Report 02-86 Military Aircraft. November, 2001. Web. Retrieved 2 March 2015
- Images, *www.af.mil*
- Images, www.*ftnews.firetrench.com*
- Images, *www.FulcrumGallery.com*
- *www.nasa.gov*
- *www.popularmechanics.com*
- Author's research and data basing of SETP membership folders, March 2015
- "The Role of the Society of Experimental Test Pilots in the Development and Testing of Unmanned Air Vehicles" SETP UAV Development and Flight Test Committee, Rogers Smith, Chairman. 19 July 2012.
- Images, *www.tarstarkas.net*

25

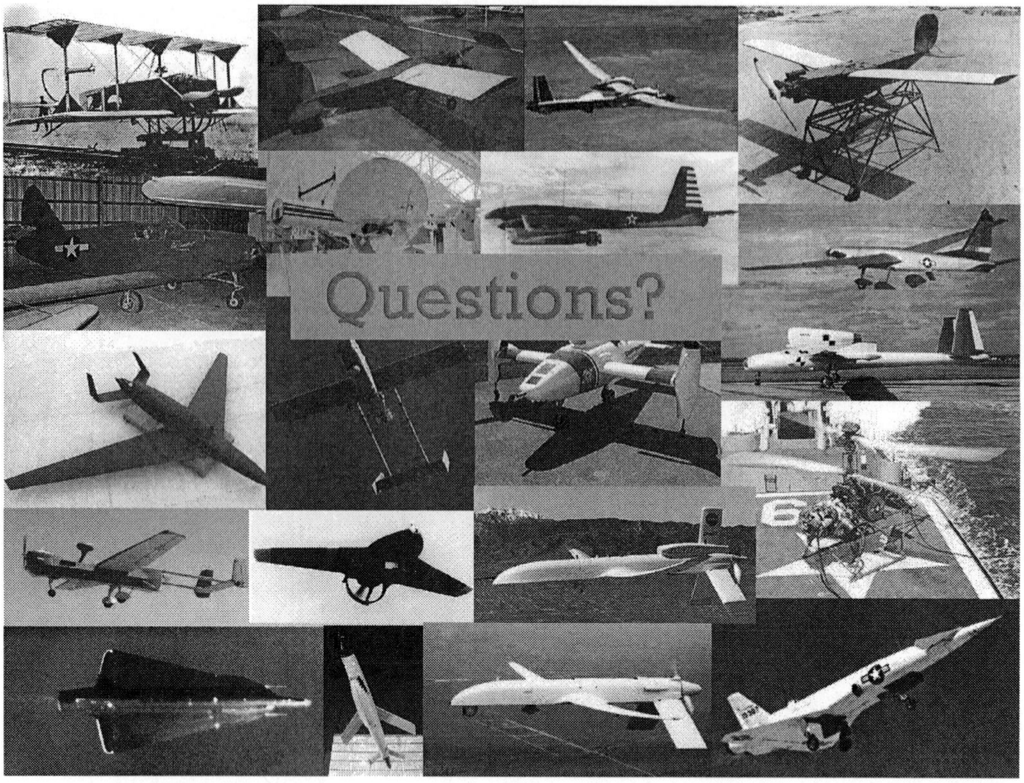

Backup Slides

28

UAS performing testing worthy of note

First Flights per Decade
of U.S. Military-Use Unmanned Aircraft
(wingspan ≥ 5 m)

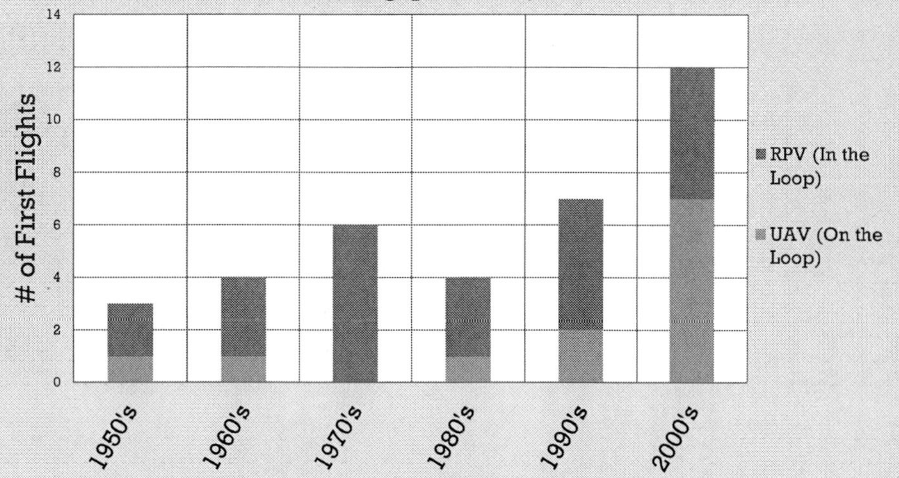

29

Recent RPV Examples: ≥ 5m Wingspan

- RQ-2 Pioneer
- GNAT 750
- RQ-5 Hunter
- RQ-1 Predator
- Eagle Eye Tiltrotor
- MQ-9 Reaper
- MQ-1C Warrior
- RQ-170 Sentinel
- X-48
- Predator C Avenger

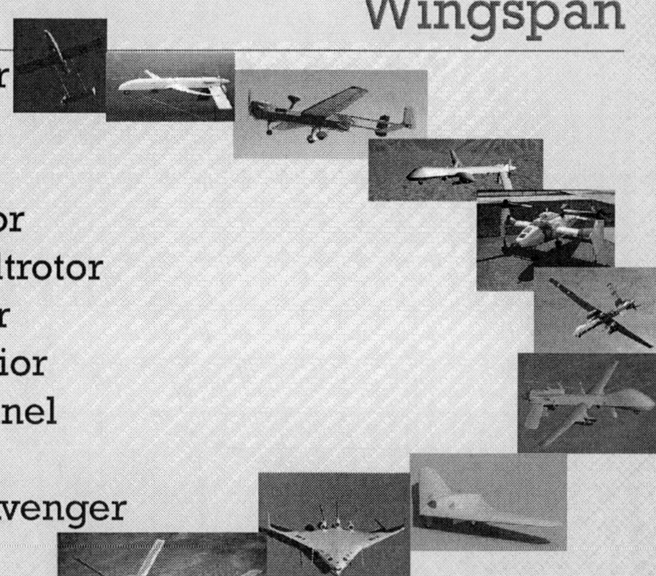

30

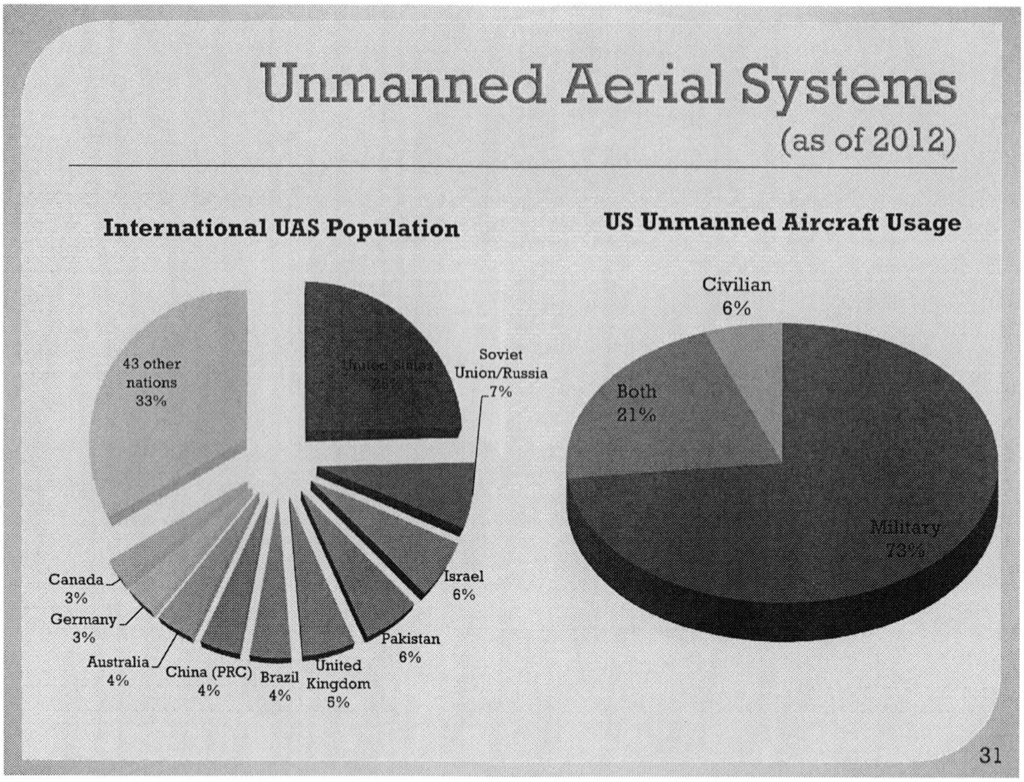
Unmanned Aerial Systems
(as of 2012)
International UAS Population
43 other nations 33%
Soviet Union/Russia 7%
Israel 6%
Pakistan 6%
United Kingdom 5%
Brazil 4%
China (PRC) 4%
Australia 4%
Germany 3%
Canada 3%
US Unmanned Aircraft Usage
Civilian 6%
Both 21%
Military 73%
31

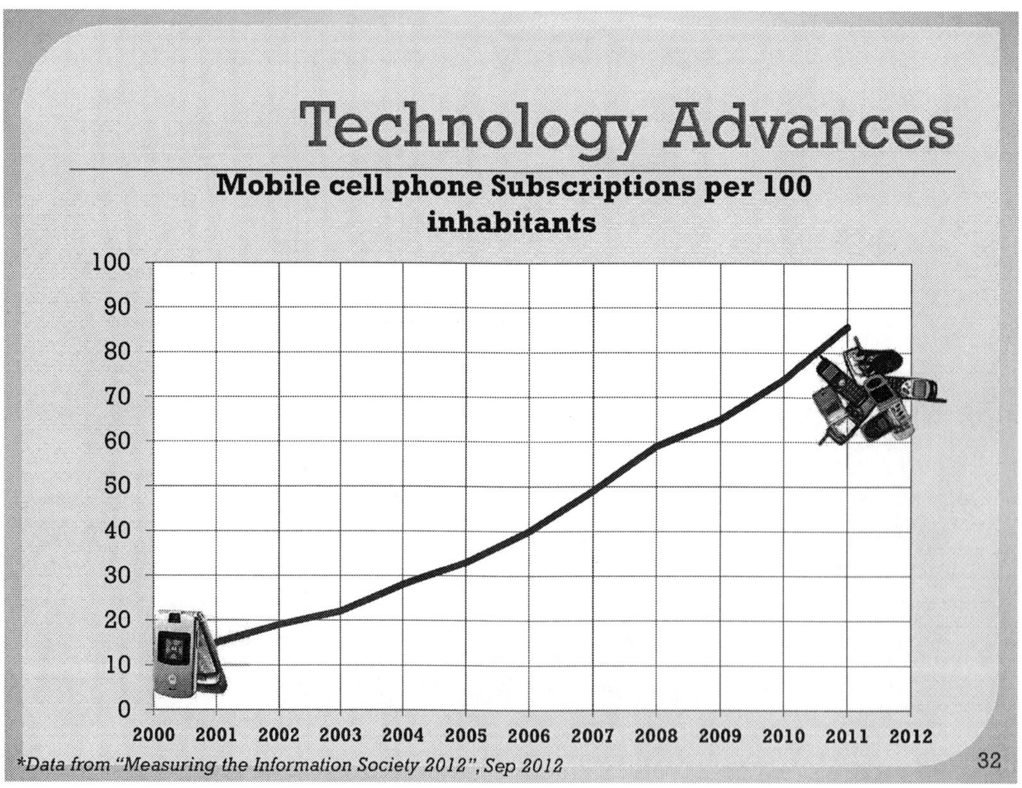
Technology Advances
Mobile cell phone Subscriptions per 100 inhabitants
100
90
80
70
60
50
40
30
20
10
0
2000 2001 2002 2003 2004 2005 2006 2007 2008 2009 2010 2011 2012
*Data from "Measuring the Information Society 2012", Sep 2012
32

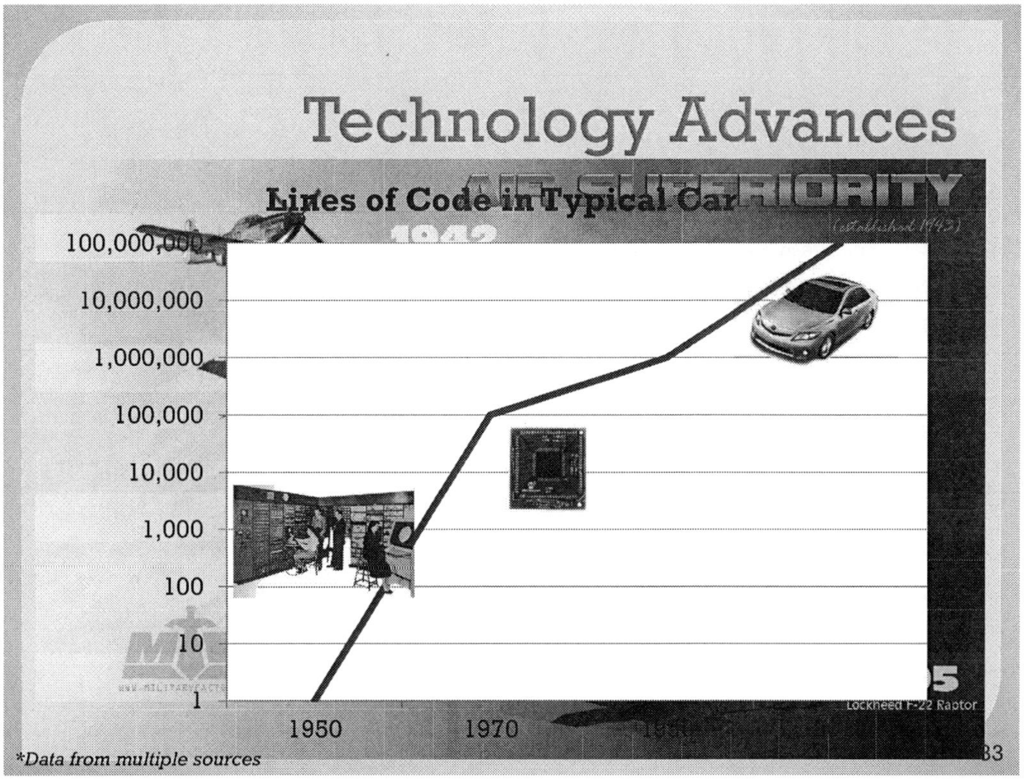
Technology Advances
Lines of Code in Typical Car
100,000,000
10,000,000
1,000,000
100,000
10,000
1,000
100
10
1
1950
1970
Lockheed F-22 Raptor
*Data from multiple sources
33

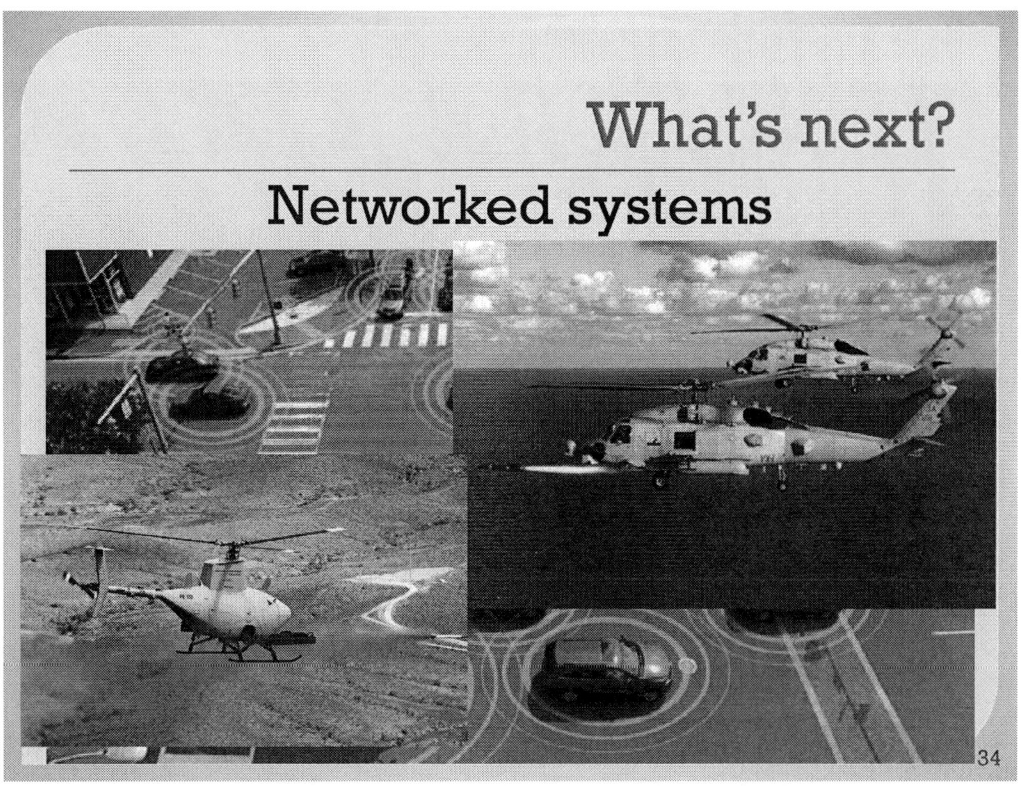
What's next?
Networked systems
34

SETP Membership Counters

- Should some UAS tests be valid membership counters?
  - Co-pilots can be Associate Members right now
- **Experimental**
  - FT under conditions not previously tested
    (1st flights, EE, Init Performance/HQ of new/sig modified a/c)
- **Developmental**
  - Initial FT of engineering/design change
    (structures, control law, systems test that assist cntl of a/c)

12 mo + 12 Flts = M
6 mo + 6 Flts = AM

- **Engineering Evals**
  - Gather previously un-established data,
    but not deemed Experimental or Developmental

12 mo = AM
6 mo +TPS = AM

- **Production**
  - 1st flight following a/c manufacture

24 mo = AM
6 mo + TPS = AM

35

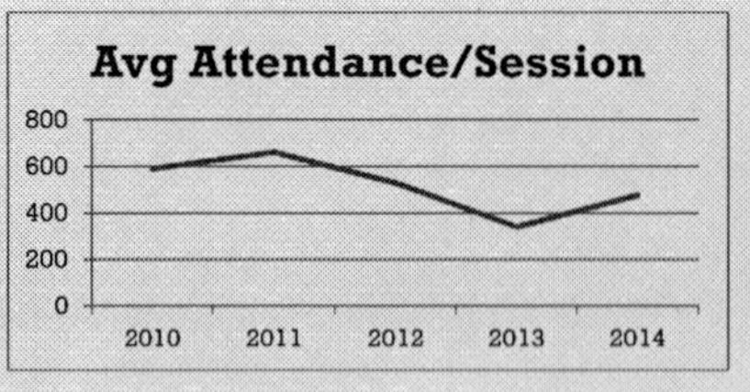

36

37

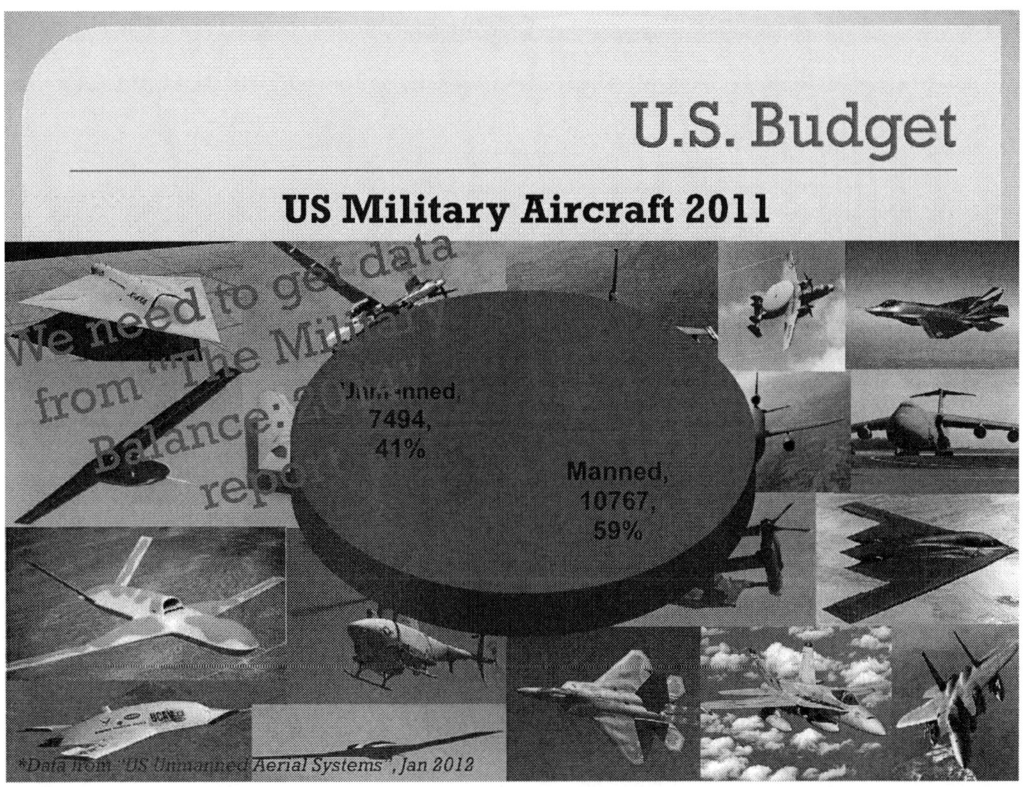
U.S. Budget
US Military Aircraft 2011
We need to get data
7494,
41%
Manned,
10767,
59%
*Data from "US Unmanned Aerial Systems", Jan 2012

Engineering, Operations & Technology
Boeing Test & Evaluation

Test & Evaluation's Role in Breaking the Development Cost Curve

Terry E. Tomeny
Director, Early Test Engagement
Boeing Test & Evaluation

Test & Evaluation's Role in Breaking the Development Cost Curve

Agenda

1) Exactly what is the Development Cost Curve?
2) Is there really a problem?
3) What is T&E's historical role?
4) What historically happens?
5) What are the real drivers?
6) What can T&E do to help?
7) Lessons learned? – or just observed?

Product Life Cycle

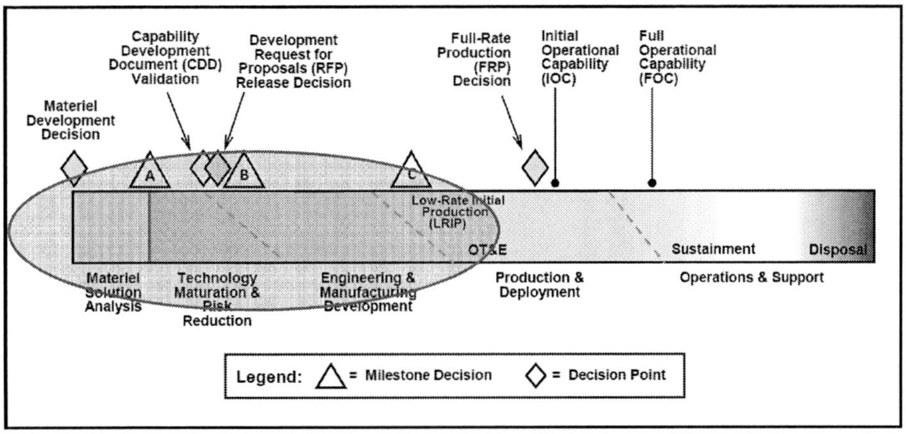

Development – Production – Sustainment - Disposal

What Really is the Cost Curve?

Augustine's Law

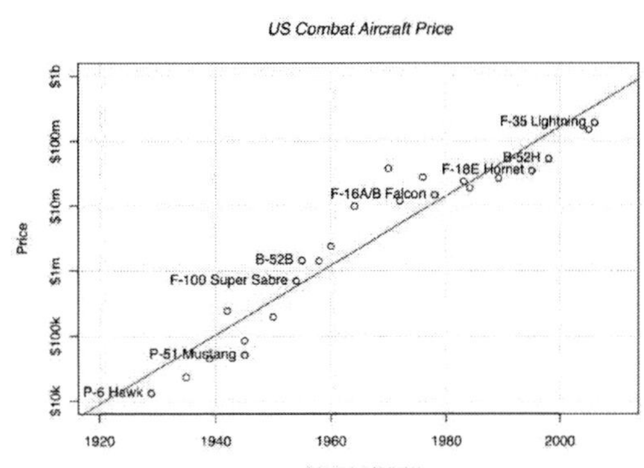

His most cited law is number 16, which shows that defense budgets grow linearly but the <u>unit cost</u> of a new military aircraft grows <u>exponentially</u>:

> "In the year 2054, the entire defense budget will purchase just one tactical aircraft. This aircraft will have to be shared by the Air Force and Navy 3½ days each per week except for leap year, when it will be made available to the Marines for the extra day."

Development Cycle Trends

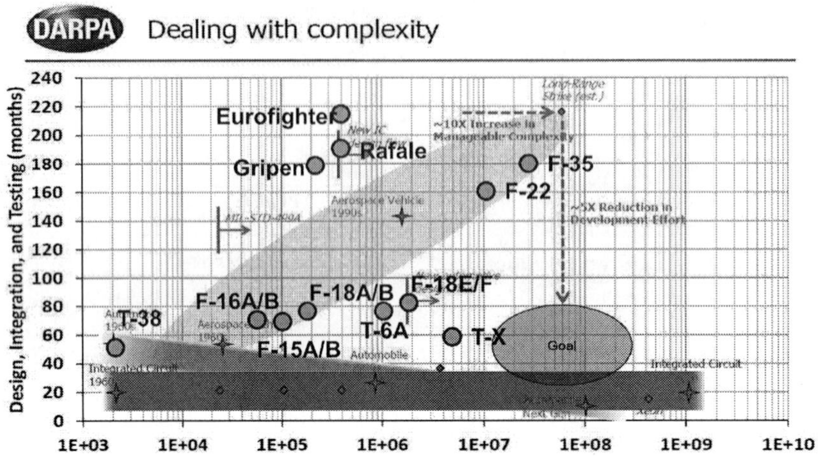

Historical Cost Growth *(in then-year dollars)*

Aerospace Systems (1960–present)	8-12%/yr
Automobiles (1960–present)	4%/yr
Integrated Circuits (1970–present)	~0%/yr

APPROVED FOR PUBLIC RELEASE. DISTRIBUTION UNLIMITED.

Engineering, Operations & Technology
Boeing Test & Evaluation | Breaking the Cost Curve

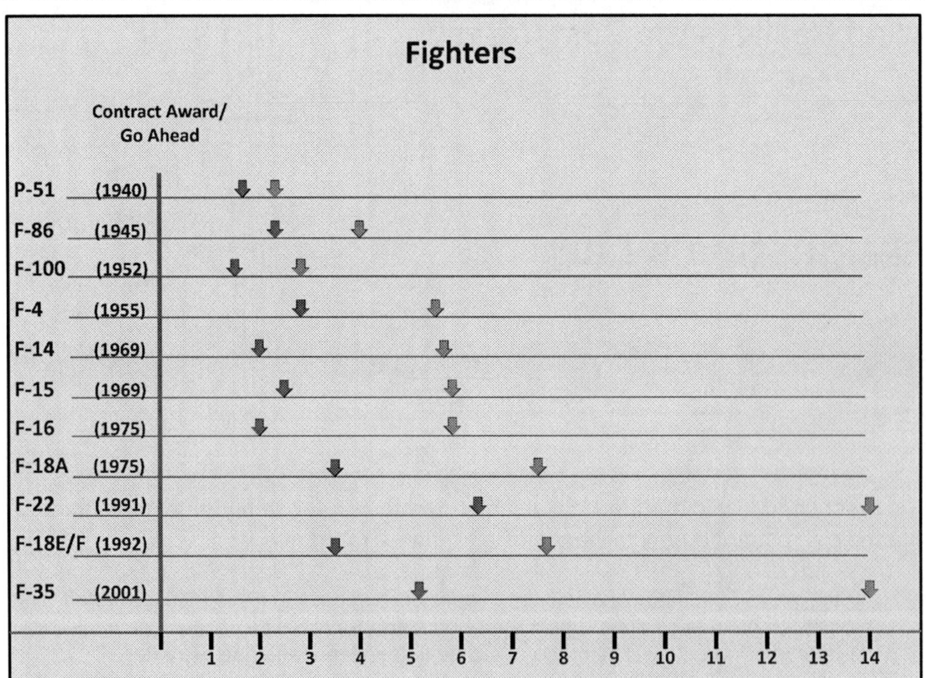

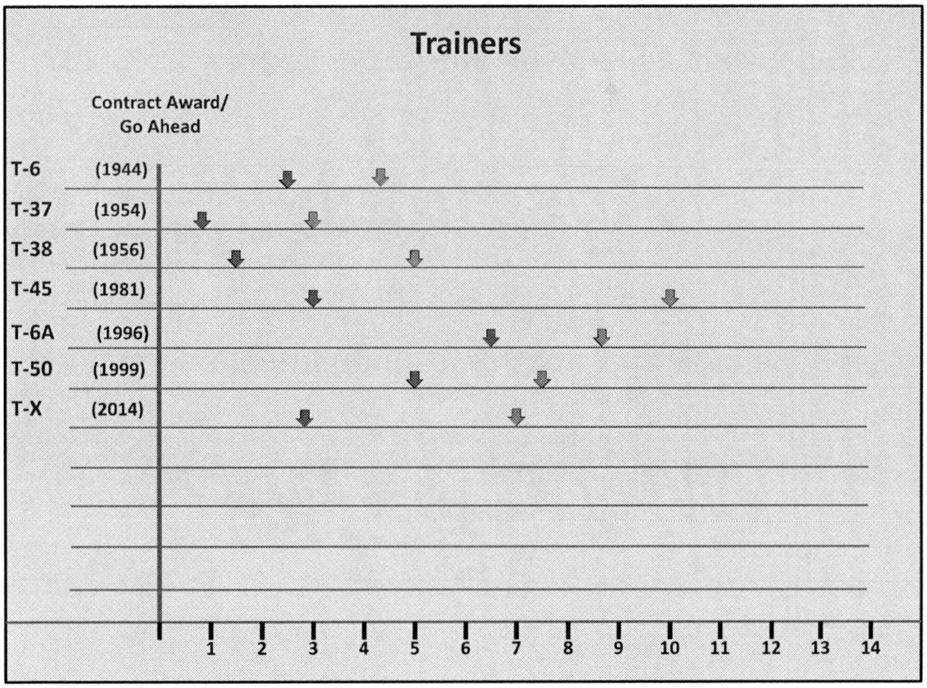
Trainers
Contract Award/
Go Ahead
T-6 (1944)
T-37 (1954)
T-38 (1956)
T-45 (1981)
T-6A (1996)
T-50 (1999)
T-X (2014)
1 2 3 4 5 6 7 8 9 10 11 12 13 14

 7

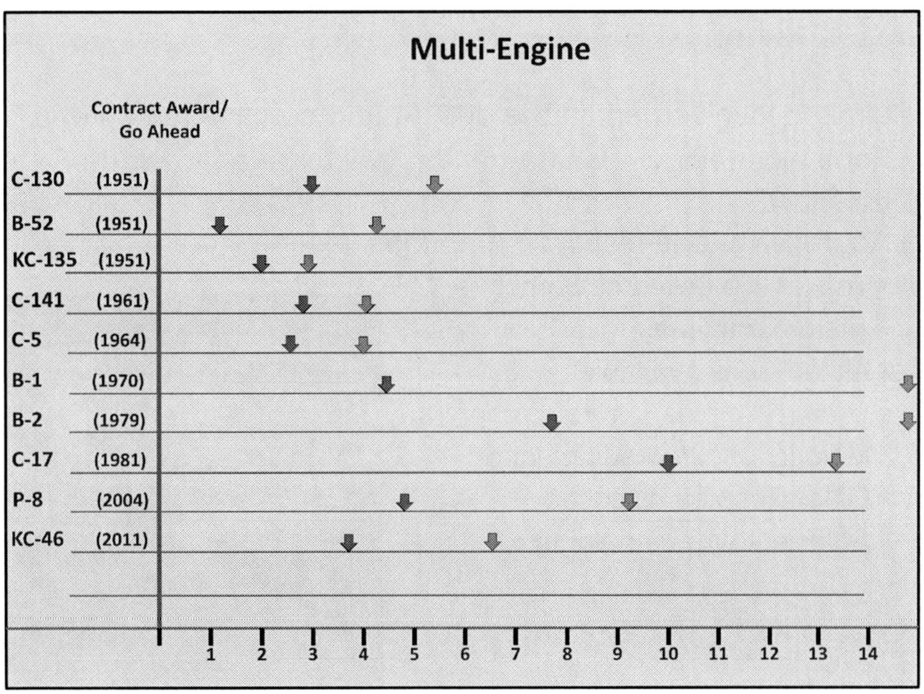
Multi-Engine
Contract Award/
Go Ahead
C-130 (1951)
B-52 (1951)
KC-135 (1951)
C-141 (1961)
C-5 (1964)
B-1 (1970)
B-2 (1979)
C-17 (1981)
P-8 (2004)
KC-46 (2011)
1 2 3 4 5 6 7 8 9 10 11 12 13 14

 8

So do we have a development cost curve problem?

Absolutely!

And if the auto and electronics industry can beat this....

Why Can't We?

T&E's Historical Role

- Limited involvement in the design and certification criteria
- Football analogy – last to get the ball to take it over the goal line
- Aggressive, optimistic assumptions
  - Test article delivered complete, mature, on time
  - All testing is available
  - Lean resources (aircraft, manning)
  - Majority of cert credit planned
  - Little expected discovery
  - Sufficient time remaining

What Commonly Happens?

The Start

- Key Players – BD, P/BM, ENG, MAN (SM), T&E
- Lean, Aggressive, Optimistic Baseline Plan
- MR Held for Late Discovery
- Phasing of Manpower
- Design Right, Build on Time, Little Discovery in Test

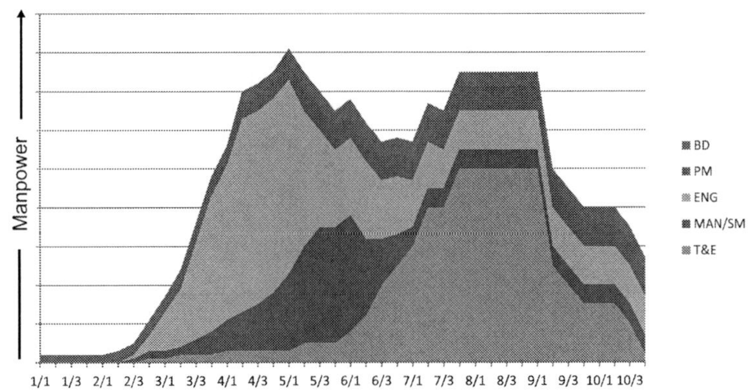

What Commonly Happens?

The First Slip

- Design Takes Longer Than Planned
- ENG Manpower Extended
- SM Complications
- MAN Starts Build with Design Changing
- Drives Longer Build

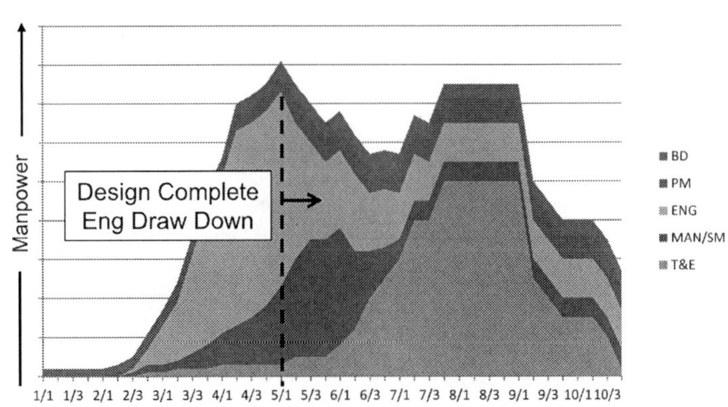

What Commonly Happens?

The Second Slip

- MAN starts build with design changing
- Delays & rework extend the build
- ENG & MAN manpower extended
- T&E ramp up begins to be ready
- Pressure grows to deliver & start testing

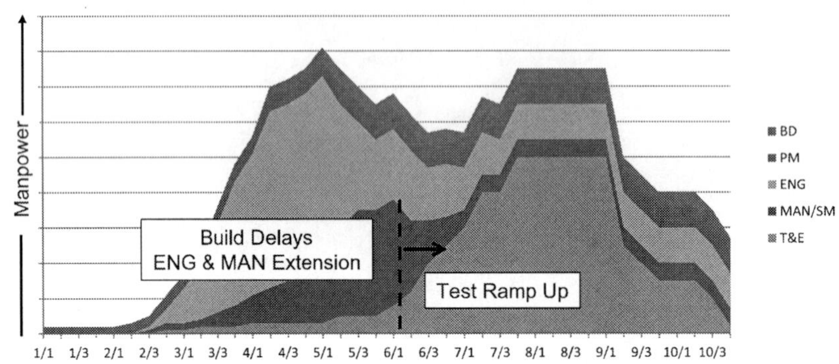

13

What Commonly Happens?

Late Delivery

- Late delivery of test articles
- Incomplete build, immature systems
- Unplanned operating limitations
- T&E "Army" ready & waiting
- Continued design changes
- Delays in reducing ENG and MAN

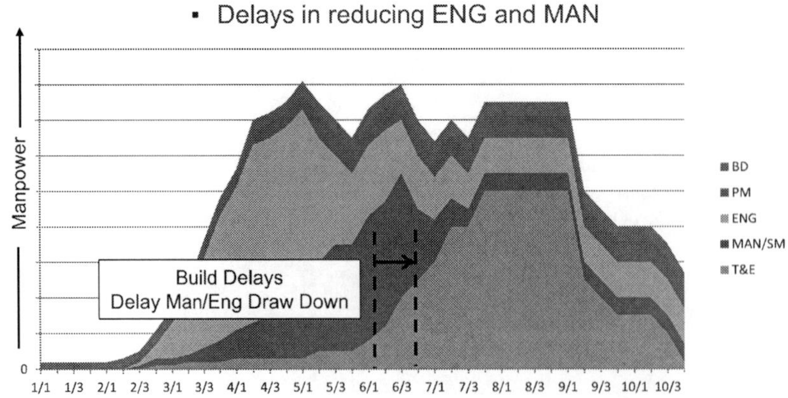

14

What Commonly Happens?

The "BULGE"

- Late start – behind the power curve
- Inefficient testing due to unplanned operating limits
- Desire to "go faster" – resource/process limited
- Addition of test resources (people, planes)

Unplanned "incident" impact - SAFETY
Unplanned discovery – analysis, redesign, mod, test
Test requirements scrub

Negotiations to Reduce Minimum Operational Requirements

"I Don't Care What it Costs, Get it Done as Quickly as Possible (Safely)"

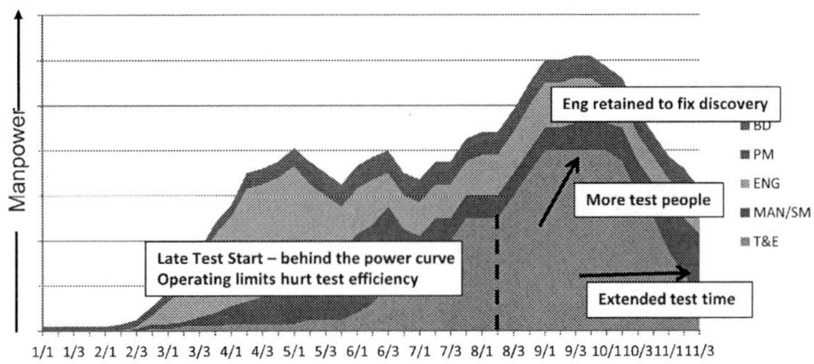

What Are the Real Drivers and Opportunities for Product Development Innovation?

<u>Design – Manufacture – Test – Produce</u>

The Seven Buckets

1. **The right people in the right place at the right time**
2. **Determining what you're going to build – the design**
3. **Determining validation & verification requirements (T&E SOW)**
4. **Determining data requirements**
5. **Determining instrumentation requirements**
6. **Integrated manufacture & test**
7. **Efficient test planning & execution**

1 The Right People in the Right Place at the Right Time

Early Engagement of Test & Evaluation

Unique Perspective

- Historically last to carry the ball in EMD
- Historically limited involvement until integrated test
- Frequent discoverer of unplanned risk

- Test & Evaluation
  - Engineering
    - Operational Experience
      - Systems Integration

Engineering, Operations & Technology
Boeing Test & Evaluation | Breaking the Cost Curve

2 Determining What You're Going to Build – The Design

- Know the requirements – get involved
- Probably the <u>biggest opportunity</u> to break the cost curve
- <u>*Design & Build the Right Thing*</u>
  - Understanding of Requirements
  - You probably know this better than anyone
- <u>*Design & Build the Thing Right – the First Time*</u>
  - First Time Quality – CVQ
  - ID risks early and do something about it
- Set Based Engineering
- Reuse
- Remember the Two Uses of Test
  - As a development tool
  - As a certification tool (V&V)

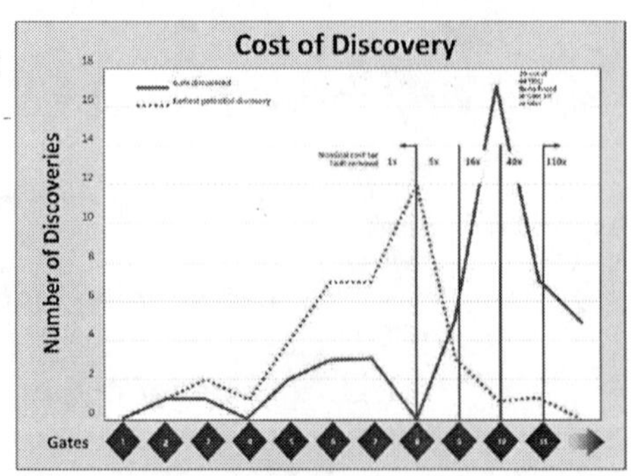

3 Determining Validation & Verification Requirements (T&E SOW)

- Get involved and understand the requirements
- Tailoring
  - "Not Applicable"
- Consciously "Shift Left"
  - Similarity
  - Inspection
  - Analysis
  - Lab (Simulation)
  - Ground Test
  - Flight Test
- Earliest and Least Expensive Certification Credit

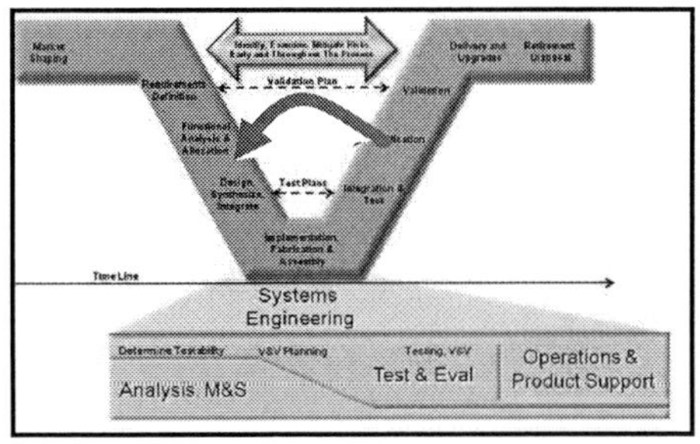

Systems Engineering "V"

4 Determining Data Requirements

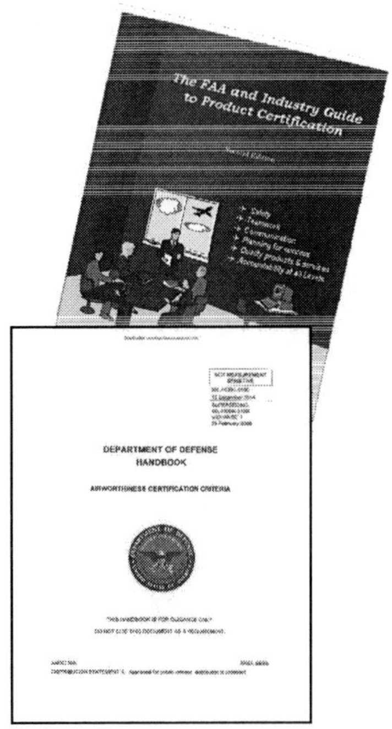

- Don't jump straight to instrumentation
- Earliest acquisition of disciplined, multiple requirements
- Sufficient data to answer the question
- Not more because we can get more
- Benefits of real-time data analysis
- Optimize where and how you get data
- Optimize data processing requirements
- Remember data analysis impact on execution

5 Determining Instrumentation Requirements

- From data – optimize instrumentation
- Stick with "tried and true"
- Significant cost driver
- Significant test efficiency driver
  - Safety of Flight
  - Safety of Test
  - Analysis Critical

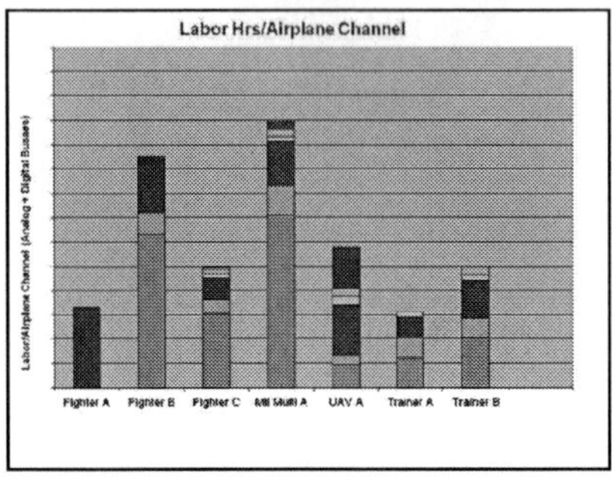

- Sufficient vs Max Instrumentation
- Consider impacts on the build process
- Lead time for suppliers

6 Integrated Manufacture & Test

- How many test articles?
- Unique configuration vs flexibility
- Test instrumentation as an aircraft system
- Integration of test instrumentation in build

- Integrate earliest test opportunities (ground test)
- Supplier management support is critical

Efficient Test Planning & Execution

- What we normally focus on
- Always remember –
  "Safety is Paramount!"
- Efficient concept of operations and operating rhythm
- Planned sortie rates, durations, and test efficiency

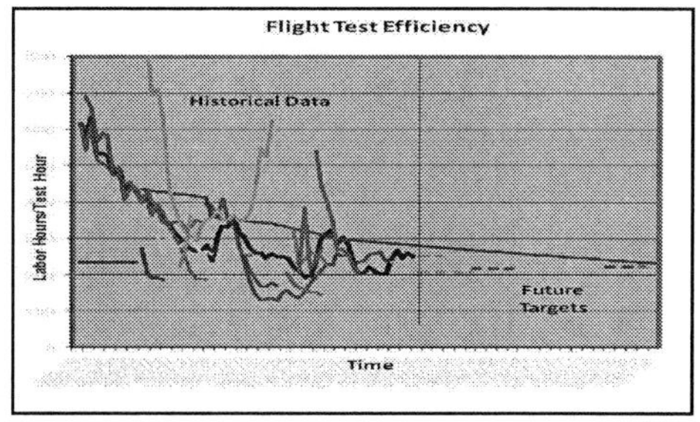

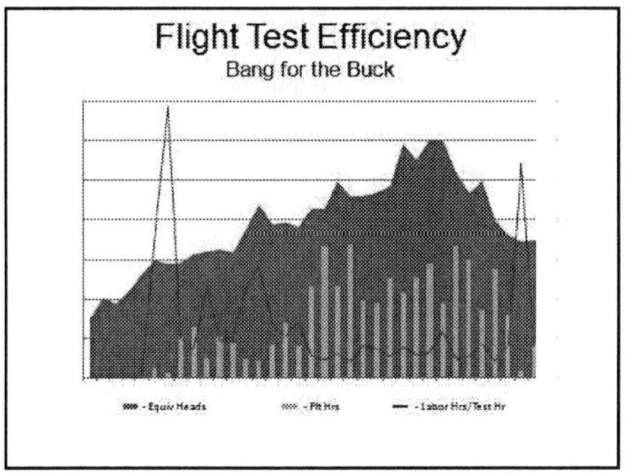

- Don't forget the "refly" factor - it's not "zero"
- Visible, easily understood metrics
- Initial resources & execution rate – with options.
- How to safely and expeditiously proceed after an unexpected occurrence
- Be ready to discuss requirements scrubs
- Be prepared to recommend deferred capability to meet IOC timelines

Lessons Learned? - Or Just Observed?

- We see the same thing recurring in almost every development program
- Get involved early – Earn a seat at the table - Value added
- It's definitely a team effort! Relationships are important.
- Too many initiatives can be distracting
  - Focus initially on the biggest bang for the buck
  - But be aware of all opportunities
- The Key is identifying risk early – and doing something about it
  - This may mean moving funding forward!
- Innovation opportunities seem to fall into seven major buckets
  - Different functional groups drive the buckets, but they can't do it alone
- Consider a periodic "report card" from program leadership
  - Program management, engineering, manufacturing, test, etc

Variable Stability Learjets – First Flights of an Old Airplane with a New VSS

Zachary A. McCarley (M), Calspan Test Pilot

John Lindak, Calspan Flight Test Engineer

Abstract

Lear 4 is the fourth variable stability system (VSS) Calspan modified Learjet. Modifications include installation of a hydraulically actuated, irreversible, digital FCS that replaces the reversible right seat controls. This VSS control system overlays the mostly unmodified reversible Learjet control system that remains connected to left seat controls. A third FCS mode, connected to right seat hydraulic controls, is the emergency fly-by-wire (FBW) mode which bypasses the VSS computer. Due to a tight schedule before the system's initial European deployment it was necessary to draw on past experience to develop and execute an efficient test plan with the goal of completing checkout of a highly modified experimental variable stability aircraft. Flight test consisted of functional checks on existing Learjet systems, safety trips, VSS envelope expansion, and feedback gain expansion. Flight test challenges included an unexpected test event found later to be nearly identical to results from a previous VSS test program; structural coupling during high feedback gain testing, and axis confusion with INS feedbacks. Successes included, safe program execution, and an on time mission capable deployment to Europe. Multiple lessons learned apply to FCS/VSS testing specifically and generally emphasize the predict-test-validate mantra.

Acronyms

AOA	Angle of attack
AOB	Angle of bank
AOS	Angle of sideslip
EGI	Embedded GPS/INS
EP	Evaluation pilot (right seat)
FBW	Fly by wire
FCF	Functional check flight
FCS	Flight control system
FTE	Flight test engineer
g	Gravitational acceleration
GPS	Global positioning system
Hz	Hertz
INS	Inertial navigation system
LL	Lesson learned
Ny	Acceleration in y-axis
Nz	Acceleration in z-axis
p	Roll rate
PSD	Power spectral density
q	Pitch rate
r	Yaw rate
SP	Safety pilot (left seat)
STC	Supplemental type certificate
TTRM	Top tail rolling moment
VSS	Variable stability system

I. Introduction

Calspan modified its fourth Learjet into a full authority variable stability platform between 2013 and 2015 to satisfy contracts for variable stability instruction at European test pilot schools – including Empire Test Pilot School (ETPS) at Boscombe Down, UK, and Ecole du Personnel Navigant d'Essai et de Reception (EPNER) at Istres, France, and Test Flying Academy of South Africa (TFASA) out of Jerez, Spain. Although Calspan had modified three previous Learjets into similar platforms, approximately a decade passed prior to the Lear 4 modification. In that time the technology and the design team had changed. The time interval between each Learjet modification caused each to be prototypical. The yearlong modification of Lear 4 was followed by flight test which took place over a period of three weeks between March and April 2015 followed by immediate deployment.

The flight test focused on achieving experimental category airworthiness and mission readiness. To assure test efficiency the flight test centered on systems that were significantly changed from previous VSS Learjets. The availability of redundant flight control systems (FCS) including the original Learjet reversible FCS was leveraged. Despite past experience this condensed flight test period identified unexpected issues and some unexpected events occurred. This paper discusses the flight test planning, execution, and lessons learned from Calspan Lear 4 VSS modification program.

II. Modifications

The primary modification to the Learjet was the removal of the right seat flight controls and replacement with a FBW type flight control system. This system employs an installed VSS computer and installed hydraulic actuators to control the airplane with the same variable stability digital FCS and an emergency analog mode as previously modified Calspan VSS Learjets. The left seat controls were unmodified from the original direct linkage system with the exception that the bob weight and aileron-to-rudder interconnect were removed. Figure 1 through Figure 4 show pre, during, and post modification.

The general modification included differences from previous Lear modifications that were expected to be of significance during the Lear 4 flight test. The automatic safety trip circuit board that is central to the safe operation of the VSS was redesigned. The control panels that the safety pilot (SP) utilizes to startup and engage the VSS were new. The inertial navigation system (INS) that is an integral part of the feedback control system was a decade newer technology and was purchased from a different manufacturer. In addition there were several other modification differences that were considered when conducting the flight test planning and execution.

Figure 1: Lear 4 Cockpit Pre-Modification

Figure 2: Lear 4 Cockpit During-Modification

Figure 3: Lear 4 Cockpit Post-Modification

Figure 4: Lear 4 Cockpit In-Flight

III. Test Planning

Test planning followed a buildup approach of starting with critical systems and expanding the aircraft capabilities in logical steps. Efficiency in the test plan was attained by viewing this modification as a buildup of four critical systems. The critical systems were the basic aircraft, the emergency fly-by-wire FCS, the safety trip system, and finally the VSS FCS. Due to the nature of the VSS Learjet modifications, proving each critical system served not only as logical buildup step but also as a layer of redundancy for failure of a subsequently tested system.

Basic Aircraft

The first logical step, was to perform a functional check flight (FCF), which included checkout of the generally unmodified ship basic systems to ensure general airworthiness. Because of extensive FCS modification this included checkout of the original Learjet reversible FCS. Completion of this checkout enabled the ability to take-off and climb to altitude with the production FCS, then engage the experimental systems up and away.

Fly-by-Wire (FBW) Checks

The second logical step after proving the basic aircraft and original Learjet FCS, was to check the analog emergency mode of operation called the “fly-by-wire” mode. While all of the modified right seat control modes are considered fly-by-wire, this secondary FCS, which is controlled by the right seat evaluation pilot (EP), bypasses the VSS FCS and safety trips and allows fixed gain control of the hydraulically actuated surfaces. It is the emergency ‘get-home mode’ should something unpleasant and mal-timed befall the left seat safety pilot (SP). This mode is critical for gaining experimental airworthiness, having removed the original control system from the right side. Completion of this checkout allows the ability to check subsequent systems with the emergency FCS available to the right seat pilot in addition to the original Learjet FCS of the left seat pilot.

The FBW system functions typically with no automatic safety trips, however for buildup it was desirable to test the system with safety trips active. Test planning to determine when the FBW mode should be flown was a chicken and egg problem with the safety trip system. It was decided that for proper buildup, the FBW system would be tested up and away without automatic safety trips to ensure emergency functionality as early in the program as possible. In place of automatic safety trips manual disconnect switches were enabled to quickly revert control to the original Learjet FCS. This testing was to be followed by testing of the automatic safety trip system to reduce the risk in landing tests of the FBW mode.

VSS Sensor Checks

As part of the buildup approach it was decided that it was necessary to check sensors in flight prior to the checking of safety trips. Instabilities in feedback control systems are as easy to program as a misplaced negative sign or crossed signals. After basic ship functionality was checked, the VSS could be powered but not engaged to allow sensor data to be collected while flying from the original Learjet FCS or from the FBW mode. Sensor signals could then be analyzed post flight to check for correct sign, scaling, and implementation. Both raw sensor data and processed signals were recorded to ensure proper implementation of the aircraft state measurements. The VSS FCS was not allowed to control the Learjet surfaces with feedback loops closed until these data had been analyzed. The following signals were to be recorded:

- All scaled inputs to the air data and complimentary filters system.
- All outputs from the air data and complimentary filters system.

- All control surface positions including engine thrusts.
- All navigation signals originating from the INS.

Safety Trip System

The third critical system was the system of safety trips that is critical to VSS safety and is the watchdog that disengages prior to flight outside of the Learjet envelope. Figure 5 shows a notional representation of the safety trip envelope related to the Learjet envelope. Safety trips of note include AOA, AOS, Nz, Ny, and top tail rolling moment (TTRM) limits. This system was a new design, though the safety limits were the same as Lear 3 and significant ground testing had been accomplished. Despite this, flight test was required to check each limit prior to testing other systems including engagement of the VSS feedback control system. Some limits were deemed acceptable by ground test due to difficulty in reaching limits in flight. For example during the Lear 3 flight test program, some primary safety trips were disabled in order to flight test secondary trips. This included the low alpha trip that required disabling the low g limit. For Lear 4, it was determined that if the low g safety trip worked in flight, and the low alpha trip worked during ground testing an adequate level of risk was mitigated. It was not desirable to increase test risk by disabling the low g safety trip in flight. Completion of this checkout allowed the ability to check functionality of the VSS FCS with a level of safety near that of the final product.

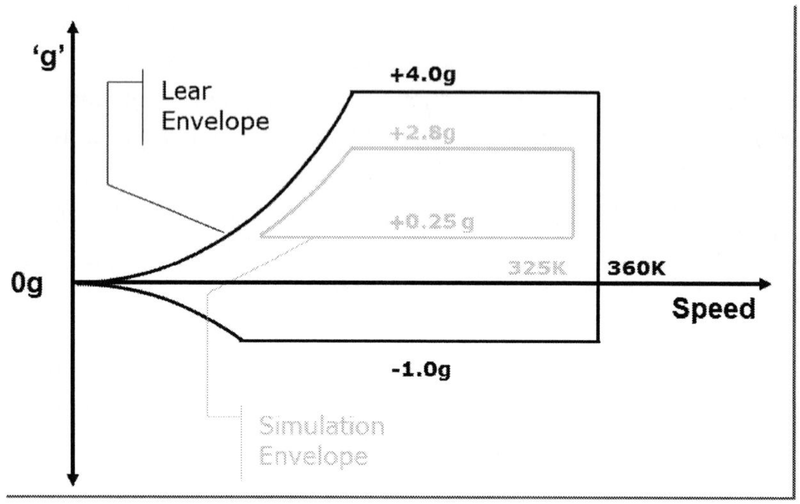

Figure 5: Safety Trip Envelope

VSS Tests

The fourth critical system was the VSS FCS which was built upon the safety afforded by the three aforementioned critical systems. This included the ability to takeoff, land, and fly entire test sorties checking only certain functionalities, as was done on the first flight to perform sensor checks. Similarly, it was determined that it was desirable to first engage the VSS without feedbacks. Step one – engage and immediately disengage. Did anything bad happen? Repeat. Allow longer periods of engagement and light maneuvering. The dual redundant FCS's again allowed this build up. The plan for VSS functionality tests including expansion of the VSS envelope, determination of feedback gain limits, and control law testing were also allowed by this dual redundancy.

VSS Envelope Expansion

The envelope of the original Learjet and its reversible FCS was well established. However, the VSS envelope had to be expanded to verify proper and safe operation throughout the simulation envelope. A spot checking of the edges of the envelope in a dynamic pressure build up approach was deemed appropriate because of the design similarity to previous VSS Learjets. The primary risk of structural coupling was mitigated via the use of an envelope expansion method of increasing q with anticipation to reduce q or disengage the system should problems arise. Control surface position and accelerometers were to be monitored for modal excitation.

VSS Feedback Gain Expansion

The feedback gain expansion was of more concern than the VSS envelope expansion because of the greater chance of structural coupling. Central to the concept of VSS operations is the ability to use a wide range of feedback gains in flight and the system's avoidance of structural coupling is critical to this functionality. The test plan called for testing gains up to the limits determined by test for Lear 3 to achieve equivalent functionality. The results of testing each feedback gain was also intended to provide an indication of the similarity of the modernized system to previous VSS Learjets.

VSS Control Laws

Lastly, the control laws for all of the different Calspan standard demonstrations had to be checked for proper function. Each is effectively a different FCS, but there were no new control laws and each had been proven on the other VSS Learjets. Therefore, it was expected that the control laws would behave themselves appropriately because of the buildup from feedback gain expansion.

IV. Test Execution

Test execution followed the buildup approach outlined in the plan, however some unique challenges were experienced and some unexpected results were attained.

First Flight

First flight was successfully conducted on 25 March 2015 with Lou Knotts as PIC and SP, Zach McCarley as SIC and EP, and John Lindak as FTE. It flew just like the simulator. More accurately, it behaved very similar to the other three VSS Learjets. Refer to Figure 6 and Figure 7.

Figure 6: First Flight

Figure 7: First Flight Complete

FBW and VSS Landings

The first FBW and VSS landing were executed during the fourth flight on 3 April 2015. The safety trips were already tested and proven. A test mode of fly-by-wire was utilize for the first landing such that safety trips were active to allow protection in case of unexpected activity close to the ground. Both landings were uneventful. Refer to Figure 8.

Figure 8: First FBW Landing

Unexpected Test Event

The trim runaway safety trip system was designed to prevent a runaway of horizontal stabilizer position that could result in a large transient upon VSS disengagement. The limit of three seconds of continuous trim disconnects the VSS before the trim reaches a level that would overstress the aircraft. This system was tested on the ground prior to flight and found to disengage the system after three seconds as intended. During flight test, this was tested again and after three seconds of trim input the system disconnected as designed. Figure 9 and Figure 10 show the flight test time history of the simulated trim runaway followed by the release of elevator hinge moment upon safety trip disconnect. This was good, but unfortunately generated unexpected results for two reasons.

First, the transient occurred earlier in the plan than expected. While performing a sign check, the SP did not see the expected indication of elevator hinge moment on the status panel (Figure 11) and continued to input the trim awaiting an indication. Post flight analysis found that the scaling of the elevator hinge moment was scaled at ten times the scaling of the other VSS Learjets. Experience with the other Calspan VSS Learjets drove pilot expectation which led the SP to hold the trim longer than planned – and prematurely into the trip. Procedures in the cards were designed for buildup but could have been more explicit.

Second, the system disconnect resulted in a larger than expected 2.8 g Nz transient. The g spike was nose-up, as shown in Figure 10. The reason that the g spike was unexpected was the failure of a lesson learned from the previous VSS test program to be heeded by the authors of the Lear 4 test plan. Figure 12 shows flight test data from Lear 3 superimposed on the Lear 4 test point showing the same transient. Clearly a lesson learned is to heed previous lessons lcarned, but it is also important to boldly document these lessons learned (LL) for future test planners. **LL – Review previous lessons learned, communicate lessons learned!**

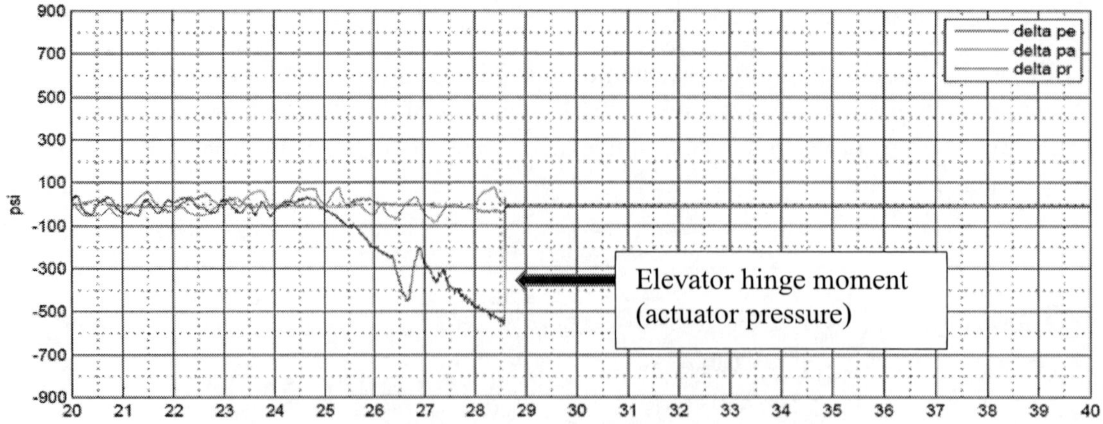

Figure 9: Trim Runaway Hinge Moment Transient

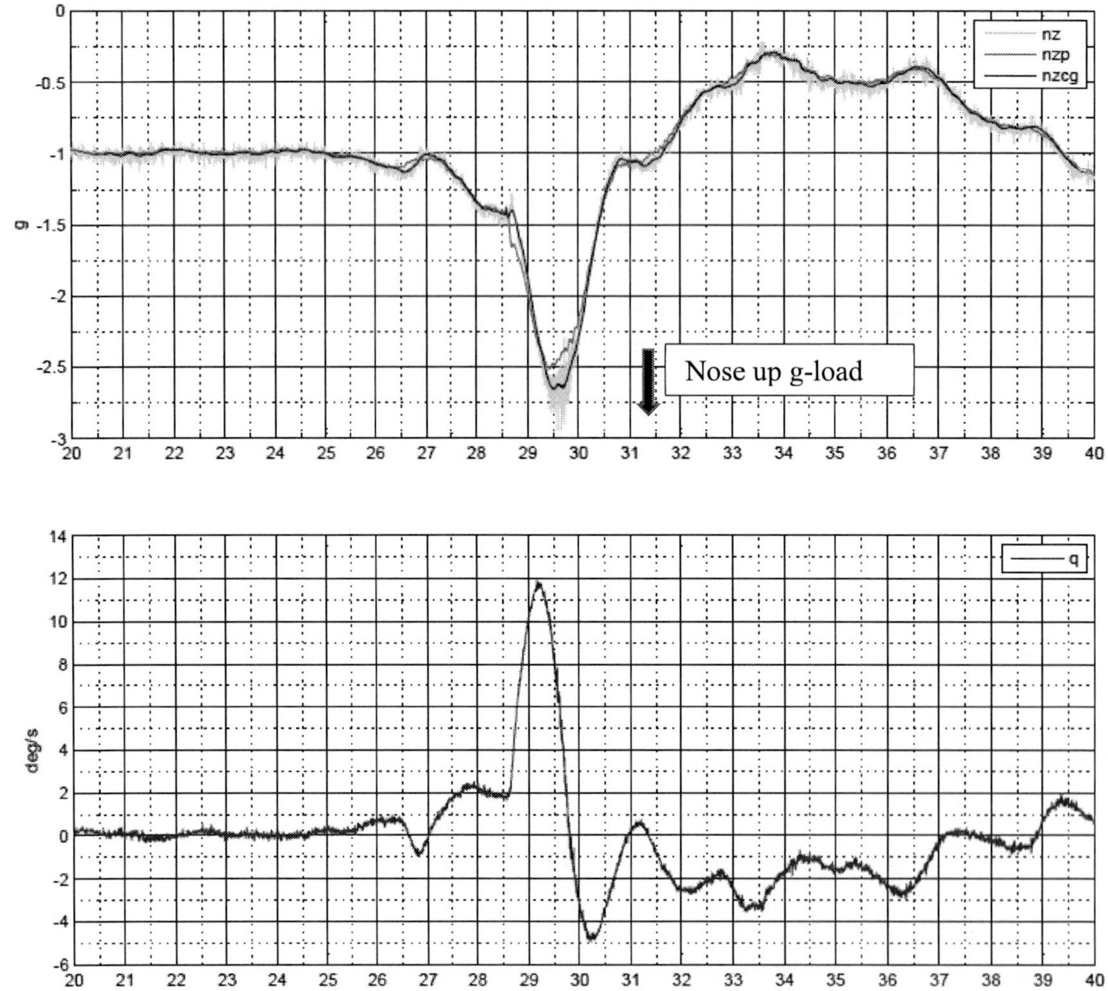

Figure 10: Trim Runaway Nz Transient

Figure 11: Status Panel Hinge Moment Indicators

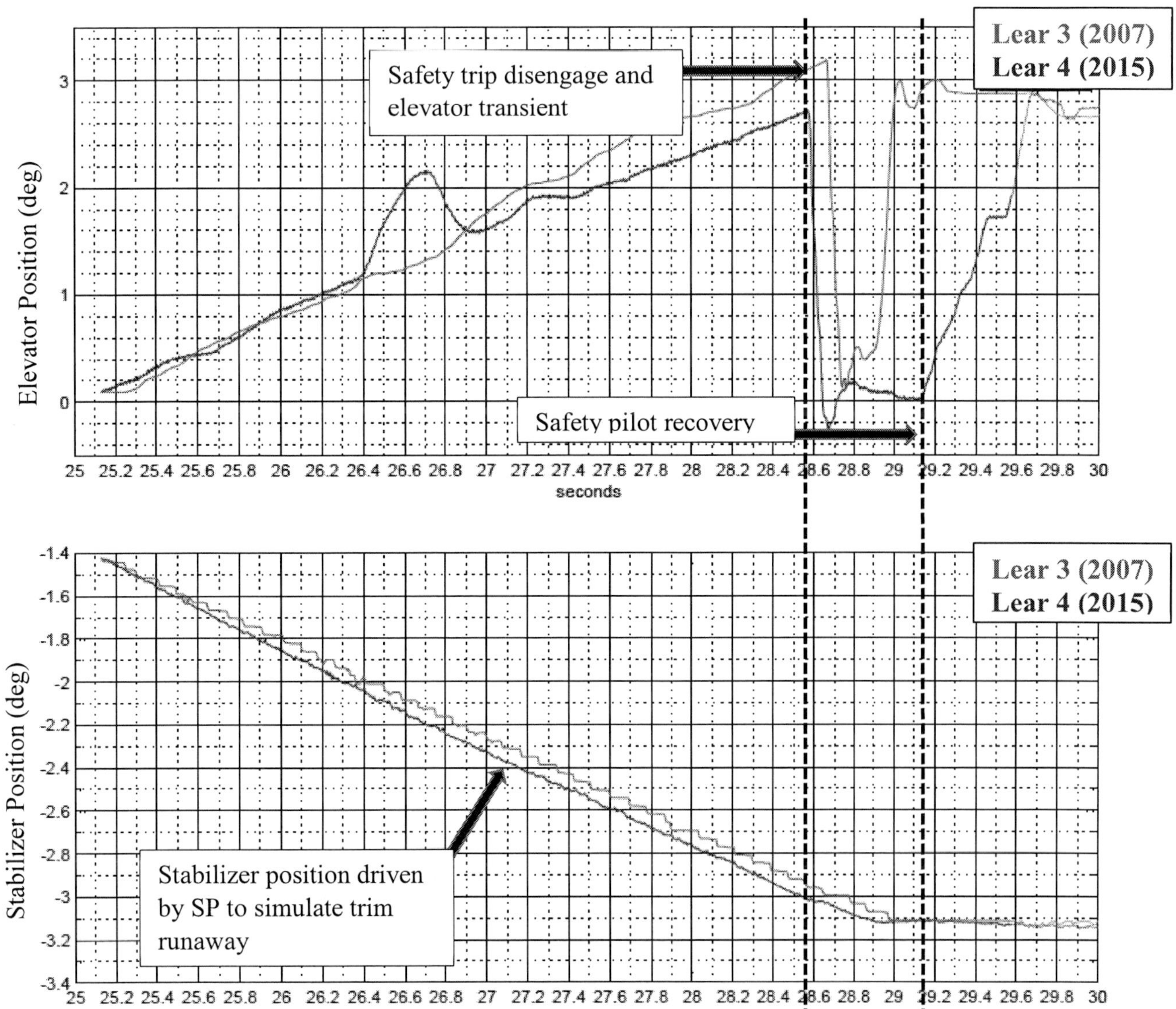

Figure 12: Lear 3 vs. Lear 4 Trim Runaway Test

Gain Testing Challenge

As discussed in the feedback expansion section, one of the critical safety concerns with VSS testing is avoiding the feedback of a signal at a frequency that couples with the aircraft structural modes. During the alpha dot feedback testing, gains were methodically increased to a level where divergent structural coupling was unexpectedly encountered. An EP elevator stick rap was applied followed by a gain increase from 1.4 to 1.6 at which time divergence occurred prior to the expected limit. Figure 13 shows the time history of divergence in addition to as PSD plots indicating that the coupling occurred at around 6 Hz. The frequency identified by the PSD plots corresponded to the 4-6 Hz Learjet symmetric wing bending mode shown in Figure 14. Upon divergence, the VSS was disengaged by the SP, to discontinue feedbacks and kill the oscillation.

This structural coupling led to the question whether an additional filter or a gain limitation was required. In reviewing current configurations and simulations it was found that a high gain alpha dot feedback was not required, so limitations were imposed. It should be noted that if a higher gain were needed, a filter to eliminate the 6 Hz coupling would be required. This additional filter would be likely to introduce undesirable delay to the FCS.

What made this test method possible was the VSS capability and pilot interface for changing gains during flight. Historically, research aircraft with FBW or augmented FCS have included the capability for pilots to adjust gains in flight, however the benefits for production aircraft test programs should also be considered. In this case, the adjustable gains allowed quick and safe determination of the gain margins for structural coupling. In the event of coupling the pilot could immediately disable feedback gains or reduce the gains to safe levels. **LL – Variable gains allows build-up with methods not otherwise available.**

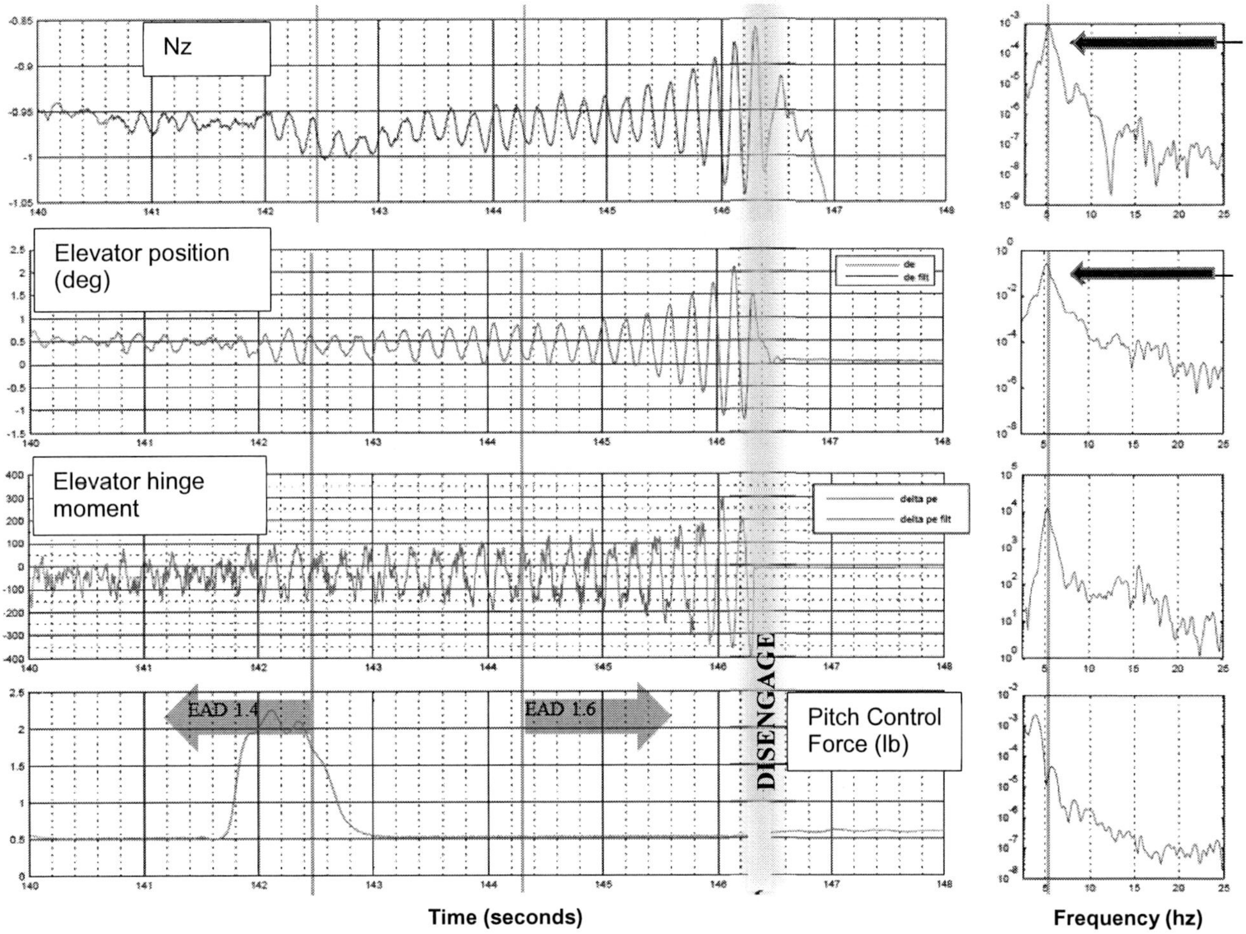

Figure 13: Divergent Structural Coupling

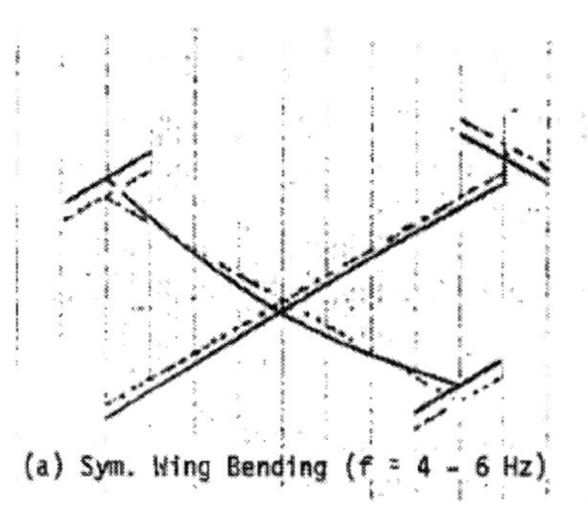

Figure 14: Learjet Symmetrical Wing Bending Mode

AXIS Confusion

The production of each Calspan VSS Learjet has been separated by about a decade. During that time technology advanced significantly, which has justified the implementation of a new INS in each VSS Learjet. Lear 3 and Lear 4 both utilize an INS that conforms to the Air Force Systems Command specification SNU 84-1 Revision D however they differ in manufacturer. The specification dictates that the I09 signal contains the "high speed vector to FCC." Following the design of the Lear3 VSS, Lear4 was setup to utilize this I09 signal for pitch, roll, and yaw rate feedbacks.

Ground testing of the Lear 4 INS included rotating the unit in all three axes to sanity check the signals. This was followed by the first flight sensor check which provided data that upon review revealed no problems with the signs or alignment of the INS axes systems. Not until late in the program during checkout of the various control laws was it found that the pitch rate command system would behave improperly at high AOB.

Initial observation of this pitch rate command FCS was that the Nz tended to increase at high AOB as opposed to the expected decrease. A review of flight data revealed that the I09 pitch rate signal from the INS was actually a theta rate signal. The difference between pitch rate vs theta rate is demonstrated in Figure 15. The lower plot indicates the bank angle of the aircraft and the upper plot is a comparison of the integrated I09 pitch rate signal and theta angle. It is expected that the signals would be equivalent at low AOB. However as AOB increased, the integrated pitch rate and theta angle should diverge. Instead, integrated theta dot (or change in pitch attitude) diverged from pitch attitude, which is nonsensical. . The roll rate (p), pitch rate (q), yaw rate (r) feedbacks were in the earth axis instead of the intended body axis resulting in the unexpected observed behavior. As AOB increased the nose would start to drift toward the ground and the FCS would attempt to maintain a constant attitude by increasing Nz.

The confusion in this case was traced back to the manufacturer's INS user manual which listed I09 signals as "pitch rate", "roll rate", and "yaw rate" which imply body axis but did not specify. That said, as long as there is an earth and machines humans attempt to control in 4D, these axis systems will continue to confuse us. We need information in both axis systems. So, sometimes the information will be in the incorrect system. Awareness of this continuing struggle is key to questioning and discovering inconsistencies. **LL – Watch your AXIS!**

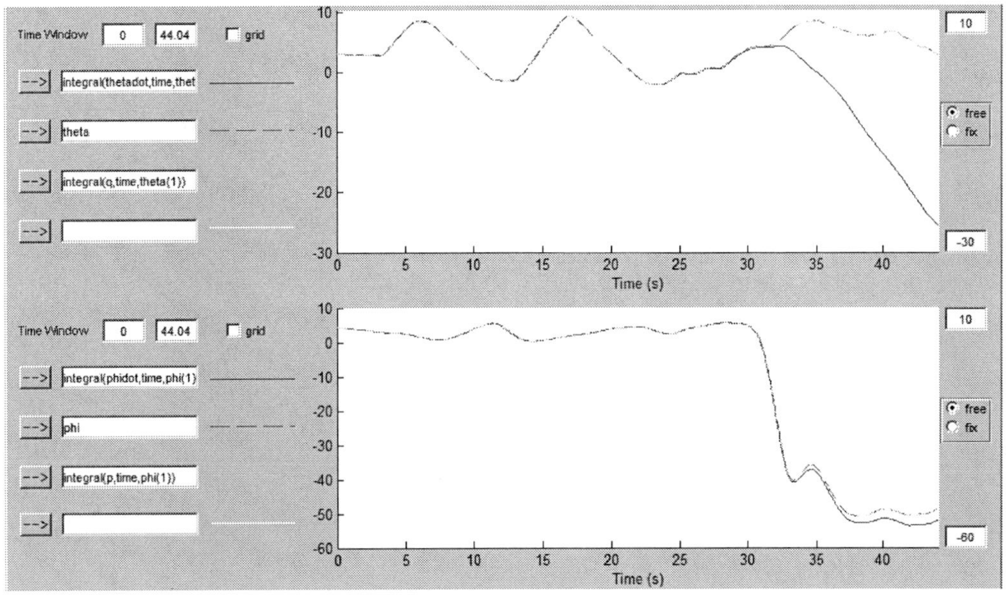

Figure 15: EGI Rate Signal Analysis

V. Lessons Learned

There were many lessons learned during this flight test program for those involved. The focus of this paper centered on three lessons:

1. Review previous lessons learned, communicate lessons learned!
2. Variable gains allow build-up with methods not otherwise available.
3. Watch your AXIS!

VI. Conclusion

The flight testing to achieve experimental category airworthiness and mission readiness revolved around the fact that the Learjet platform is an old and well proven aircraft while the installed Calspan VSS, is modernized but similar to robust and previously proven systems. During the condensed flight test period, multiple interesting questions arose and events occurred. The flight test discoveries were addressed and less than one month after the completion of the first flight, the aircraft was mission capable and successfully deployed to Europe, helping train the next generation of test pilots and engineers

AUTHOR INDEX